Sir John Budd Phear

Elementary Hydrostatics

With numerous examples. Fourth Edition

Sir John Budd Phear

Elementary Hydrostatics
With numerous examples. Fourth Edition

ISBN/EAN: 9783337075620

Printed in Europe, USA, Canada, Australia, Japan

Cover: Foto ©berggeist007 / pixelio.de

More available books at **www.hansebooks.com**

ELEMENTARY

HYDROSTATICS.

WITH NUMEROUS EXAMPLES.

BY

J. B. PHEAR, M.A.,

FELLOW AND LATE ASSISTANT TUTOR OF CLARE
COLLEGE, CAMBRIDGE.

FOURTH EDITION.

London and Cambridge.

MACMILLAN AND CO.

1866.

Advertisement to the Second Edition.

THIS Edition has been carefully revised throughout, and many new Illustrations and Examples added, which it is hoped will increase its usefulness to Students at the Universities and in Schools.

In accordance with suggestions from many engaged in tuition, Answers to all the Examples have been given at the end of the book.

The Author wishes to acknowledge the assistance given him by his brother, the Rev. S. G. Phear, Fellow of Emmanuel College, in reading the proof sheets, adding many Examples, and furnishing the Solutions and Answers.

Dec. 1st, 1857.

CONTENTS.

SECTION I.

SECTION II.

SECTION III.

SECTION IV.

SECTION V.

HYDROSTATICS.

SECTION I.

1. DEF. A *fluid* is a collection of material particles so situated in contact with each other as to form a continuous mass, and such that the application of the slightest possible force to any one of them is sufficient to displace it from its position relative to the rest.

That part of *Statics*, where a fluid appears as the principal means of transmission of force, is termed *Hydrostatics*. The law of that transmission must, like the law of transmission by a rigid body, by a free rod or string, or by contact of surfaces, &c., be established by experiment.

The mutual forces called into action by the contact of surfaces are in Statics called *pressures:* this term is used in the same sense in Hydrostatics, where it is applied to denote the forces of resistance, which adjacent particles of the fluid exert, either upon one another, or upon rigid surfaces in contact with them. The nature of the reaction between a rigid surface and a fluid in contact with it might perhaps be arrived at by the aid of analysis from the above definition. But such an investigation, even if entirely satisfactory in itself, would

1

be out of place in this treatise. It may here therefore be taken as the result of experiment that:

When a fluid rests in contact with a rigid body, a mutual force of resistance is called into action at every point of the common surface of contact, the direction of which force is normal to that surface.

2. If in the side of a vessel, containing fluid upon which forces are acting, a piston be placed, the pressure exerted upon it by the fluid particles with which it is in contact, would thrust it out, unless a force sufficient to counteract this pressure were applied to the back: this counteracting force is of course exactly the measure of the pressure of the fluid upon the piston. It is not difficult to conceive that, generally, the magnitude of this pressure would be different for different positions of the piston in the sides of the vessel; inasmuch as the portions of the fluid which it would touch at those different places, would not necessarily be similarly circumstanced, and would not therefore require for the maintenance of their equilibrium that the piston should exert the same force upon them: when, however, the pressure for *every* such supposed position of the piston, wherever taken, is the same, the fluid is said to press *uniformly;* and when not so, its pressure is said to be not uniform.

Again, it is clear that the pressure upon the piston in any given position must vary with the magnitude of its surface, and if this were reduced to a mathematical point the pressure *upon* it would be, strictly speaking, absolutely nothing, because the surface pressed is nothing; but even in this case the conception of the pressure *at* the point is perfectly definite; it signifies the capability or tendency which the fluid there has to press, and which, if existing over a definite area, would produce a definite pressure; and this view of it leads us to the following usual definition of its measure.

The pressure at any point of a fluid is measured by the pressure which would be produced upon a unit of surface, if the whole of that unit were pressed uniformly with a pressure equal to that which it is proposed to measure.

3. It is usual to represent this measure of the pressure *at* a point by the general symbol p; and whenever it is said that a surface, in contact with a fluid and containing A units of area, is pressed uniformly with a pressure p, it is meant that the pressure of the fluid *at* every point of it, measured as above defined, is equal to p units of force : hence if P be the pressure which the fluid exerts *upon* this surface A, since the pressure is *uniform*, and therefore the actual pressure *upon* each unit is p, the pressure upon the A units must be A times p, or $P = pA$.

It may be here remarked, that as P is of four linear dimensions, being the measure of a moving force, and A is of two, therefore p must be of two dimensions, *i. e.* if the linear unit be supposed increased n fold, the numerical value of p for a given pressure will be increased n^2 fold.

4. In the foregoing explanation of the meaning of the term "pressure at a point" in a fluid, the point has been assumed to be in contact with a rigid surface, which was supposed to be the subject of the pressure; now if we consider any portion of fluid, within a larger mass and forming part of it, no force but that of resistance can be exerted upon it by the surrounding fluid; for we may imagine it to be isolated from the rest by an excessively thin enveloping film, which will manifestly produce no disturbance among the particles of either portion of the fluid, because its existence neither introduces new forces nor destroys any of those which

are acting; further, this film may be supposed rigid, without affecting the relative positions or equilibrium of the particles forming the interior and exterior portions of fluid : but under this hypothesis the pressure at any point, of either the interior or exterior fluid, which is in contact with the rigid film, acquires the meaning given above, and as the introduction of the film in no way alters the actions of the portions of fluid upon one another, we thus arrive at the conclusion that different portions of a mass of uniform fluid only press against each other in the same way as they would against rigid surfaces of the same form, and therefore the term "pressure at a point" means the same thing whether the point be within a fluid or be in one of its bounding surfaces.*

5.   This last conclusion with regard to the action of different portions of the same fluid upon one another, which is of considerable importance in the solution of hydrostatical problems, does not rest solely upon reasoning analogous to that just given.   It may be considered as a fact deduced from experiment, in the same way as all other physical laws (Art. 7), that:

The statical action of any one portion of a fluid upon that which adjoins it, is the same as if the latter portion were a rigid body having the imaginary surface, which divides the two portions, as its surface of contact with the fluid.

We are therefore justified, whenever it concerns us to investigate the pressures exerted by a surrounding fluid upon an included portion, in replacing this portion by a conterminous solid.   It is generally convenient to take for such a purpose

---

* The analogy between "pressure at a point" in a fluid, and "velocity at any instant" of a moving particle, and between their respective measures, is too striking to escape the notice of the student; both terms are abstractions employed for the purpose of avoiding the constant use of the periphrasis, which is given once for all in the definitions of their measures.

the solid which would be formed by supposing the constituent particles of the portion of fluid, which it is wanted to replace, to become by any means rigidly fixed in their *relative* positions and to be still affected by the same external forces as before: it is manifest that such a solid could not differ from the fluid which it replaces, as regards its action upon the surrounding fluid, for it would itself be identical with it in every way, were it not for the circumstance that its particles are supposed to be provided with an artificial check against moving from their relative positions, in addition to, or rather instead of, the mutual resistances which effect the same end in the fluid state.

6. If a surface opposed to a fluid be itself rough or capable of exerting friction, the particles of fluid adjacent to it would, as it were, adhere to it and thus form a sort of polished veneer, because the definition of a fluid, which states that the application of the slightest force is capable of displacing the particles, precludes all idea of the existence of any tangential action between the particles themselves; and therefore although there may be resistance to the tangential motion of the particles in contact with the surface, there can be none between them and their next neighbours. For the same reason, whenever a portion of fluid is supposed to become solidified, its surfaces must be considered perfectly smooth. Hence in all cases the pressure of fluids is normal to the surfaces pressed. (Art. 1.)

It is true that very few fluids answer strictly to the definition given above (Art. 1); there is generally a certain amount of friction or viscosity between their particles, and in all cases, where the instantaneous effect of forces upon a fluid is the subject of investigation, this mutual tangential action between the particles cannot be neglected. But it is found practically that when once equilibrium is established, the particles have always assumed such a position *inter se* that no tangential action is called forth; and therefore it is immaterial to consider whether the capability for it exists or not. Thus if any semi-fluid such as honey or treacle be allowed to assume its position of equilibrium under the action of gravity, it will do so very slowly compared with water under the same circumstances, but in the end it will be found that its position is exactly the same as that of the water. Hence in *Hydrostatics* all fluids whatever may be assumed to be perfect fluids.

7. The law of transmission of pressure through fluids, which was alluded to above, may be stated as follows:

A force applied to the surface of a fluid at rest is transmitted, unchanged in intensity, in all directions through the fluid.

Like all other physical laws, this is experimental; or rather it is suggested by experiment, and its proof is deduced from a comparison of the results of calculation based upon it, with those of corresponding observations: in this sense the following experiment may be said to prove it.

The annexed figure represents a vessel of any shape containing a fluid, which may be supposed to be acted upon by gravity, as must generally be the case, or by any other forces whatever: into the sides of this vessel are fitted any number of pistons, represented by $A_1, A_2...A_n$, and having plane faces whose areas are respectively $\alpha_1, \alpha_2...\alpha_n$; sufficient forces are also supposed to be applied to these pistons to keep them in their places; in fact whatever be the forces whether only gravity or

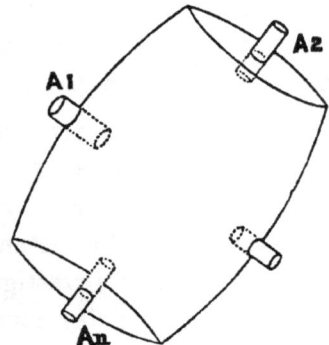

any thing else which are acting upon the system, the whole is supposed to be in equilibrium. If now any additional force as P_1 be applied to the piston A_1, it is found that to preserve equilibrium additional forces P_2, P_3...P_n must be also applied to the pistons A_2, A_3...A_n respectively such in magnitude that $\dfrac{P_1}{\alpha_1} = \dfrac{P_2}{\alpha_2} = \&c. = \dfrac{P_n}{\alpha_n}$: this result clearly shews that the application of a pressure $\dfrac{P_1}{\alpha_1}$ upon each unit of area of the piston A_1 has caused the same additional pressure upon every unit of area in each of the other pistons. In this way may the truth of the principle enunciated be verified.

Cor. It follows from this that the pressure at any point *within* a fluid mass is the same for all directions. For the action between any two adjacent portions of the fluid at any point would be the same as would exist if we suppose one of the portions to become rigid (Art. 5). In this case the pressure at the point would be caused by a rigid surface pressing on the fluid, and therefore, by the principle just enunciated, would be the same in all directions.*

8. Of fluids there are some, such as air, whose volumes or dimensions are increased by diminishing the pressure upon them and *vice versa;* these are commonly called *elastic* fluids, and all others *inelastic.* Inelastic fluids are also often distinguished by the name of *liquids,* while elastic are called either *gases* or *vapours* according as their state is one of permanent or temporary elasticity. It is probable that every fluid is compressible, when very great pressure is employed for the purpose, although within the limits of the forces with which we shall be concerned no appreciable error will be

---

* The fact of fluid pressure at a point being equal in all directions, leads immediately to this: that the *resultant* pressure upon *any* indefinitely small surface passing through that point must be normal to the surface. (Art. i.)

committed by considering water, mercury, &c. which con-
stitute the inelastic fluids, or liquids, as incompressible.

9. The conception of the *mass* of any portion of a fluid
is the same as that of a solid, and its measure is also the
same: thus if M be the mass of a portion of fluid whose
weight is W, the accelerating force of gravity being g, we
have the relation,

$$W = Mg\ldots\ldots\ldots\ldots\ldots(1).$$

10. If any equal volumes of a fluid, wherever taken
throughout its extent, always contain the same mass, the
fluid is said to be of *uniform density* or *homogeneous:* otherwise its density is not uniform; it may evidently vary by
insensible degrees from point to point.

The density *at* any point is measured by the ratio between
the mass which would be contained in a unit of volume, if
the fluid throughout that volume were of the same density as
at the proposed point, and the mass contained in a unit of
volume of some homogeneous standard substance. Thus if
V be the volume of the mass M in the previous example, and
if the density of the mass be the same *at every point*, and be
represented by ρ, the mass in each unit of the volume V will
be $= \rho$, and $\therefore M = V\rho$, the unit in which this mass is
estimated being the mass of a unit of volume of the standard
substance. We might write therefore, instead of the above
form,

$$W = \rho V g\ldots\ldots\ldots\ldots\ldots(2).$$

NOTE.—It is very important to remember the unit of
weight in terms of which W is here given.

It can be discovered as follows:

The equation

$$W = \rho V g$$

signifies that the weight of the volume V of any substance of which the density is measured by ρ, equals $\rho V g$ times the unit of weight;

$$\therefore \text{ putting } V = 1, \ \rho = 1,$$

the weight of the volume unity of any substance, of which the density is unity, equals g times the unit of weight.

Hence, *the unit of weight assumed in formula* (2) *equals the* g^{th} *part of the weight of the unit of volume of the substance by reference to which ρ is estimated.*

11. It is sometimes convenient, in reference to uniform fluids, to consider the *weight* of the portion contained in a unit of volume, rather than the *mass* of the same portion as we do when we speak of density; the term used to designate the particular quality thus referred to is *specific gravity*, which is generally defined as follows:

The specific gravity of a uniform fluid is measured by the weight of a unit of its volume estimated in terms of THE WEIGHT OF A UNIT OF VOLUME *of some particular standard fluid, taken as* THE UNIT OF WEIGHT; *i. e.* the specific gravity of any substance is the ratio between the weight of any given volume of it and the weight of the same volume of the standard substance.

If therefore V and W denote the same things as in the preceding examples, and S be the specific gravity of the fluid, we get

$$W = SV \ldots \ldots \ldots \ldots \ldots (3).$$

NOTE.—By expressing in words the meaning of this equation, it appears very clearly, as in (Art. 10, Note), what is the unit of weight here assumed.

The equation (3) declares that

the weight of a volume V of any substance of which the specific gravity is S, equals SV times the unit of weight;

$$\therefore \text{ putting } V = 1, \ S = 1,$$

it follows, that

the weight of a volume unity of a substance of which the specific gravity is unity equals the unit of weight;

or the assumed unit, in terms of which W is given in equation (3), is *the weight of the unit of volume of the substance by reference to which S is estimated.*

It may be well to caution against the following error, which results from forgetting that the units of weight in (2) and (3) are different.

It is frequently concluded that, because (2) gives

$$W = g\rho V,$$

and (3) gives $\quad\quad W = SV,$

$\therefore$ *numerically* $\quad\quad g\rho V = SV, \text{ and } g\rho = S.$

The true numerical statement is

$g\rho V$ times the unit of weight in (2) $= SV$ times the unit of weight in (3);

whence we get, if the substances to which ρ and S refer be the same, $\rho = S$;

a result clearly consistent with the definitions of ρ and S: for, since weight varies as mass, the ratio between the masses of two substances equals the ratio between their weights.

12.   In order to reduce the weight of a given substance, determined by (2) or (3), to pounds or ounces, it is necessary to know in pounds or ounces the weight of the unit of volume of the standard substance.

If, as is generally the case, distilled water be the substance to which specific gravities and densities refer, and a cubit foot be the unit of volume, the unit of weight in (3) nearly equals $\dfrac{1000}{16}$ lbs., because the weight of one cubic foot of distilled water is nearly 1000 oz.

The unit of weight in (2) will then be $\dfrac{1000}{g}$ oz., and therefore is still arbitrary as long as the unit of time is arbitrary, for upon this, as well as upon the unit of length, the numerical value of g depends.

EXAMPLES TO ARTICLES 10, 11, 12.

(1) If one second be the unit of time, what must be the unit of length in order that the formula $W = g\rho V$ may give the weight in lbs., supposing the unit of volume of the standard substance to weigh 16 lbs.?

(2) Determine, approximately, the unit of time that the unit assumed in $W = g\rho V$ may equal five ounces when one foot is the unit of length and water the standard substance.

(3) Find the unit of time, when two feet is the unit of length, in order that the units of weight in $g\rho V$ and SV may be equal.

(4) Obtain the specific gravity of the standard substance referred to water, when $W = SV$ gives the weight in pounds.

SECTION II.

INELASTIC FLUIDS.

13. *The pressure at any point below the surface of a uniform fluid, which is at rest under the action of gravity alone, varies as the vertical distance below the surface.*

[DEFINITIONS. The *vertical* line at any given place is the direction of gravity at that place. It can, therefore, be practically defined as the direction of a plumb-line at rest under the action of gravity only.

The *horizontal* plane at a given place is the plane to which the vertical line at the same place is perpendicular.]

Let P be the point below the surface, p the pressure at that point, measured as in Art. (2), M the point where a vertical through P meets the surface: let MP be represented by z, and let a prism of fluid of very small transverse section α, having MP for its axis, be considered. No circumstances affecting the equilibrium of the particles of fluid will be introduced by supposing those forming the prism to be solidified into one mass, (Art. 5).

But under this supposition the rigid prism MP is in equilibrium under the action of

        its own weight vertically downwards,

        the pressure of the fluid vertically upwards upon its base α, and the pressures upon its sides which, being normal to these sides, (Arts. 5 and 1), are all perpendicular to the axis and therefore horizontal.

These two systems of vertical and horizontal forces must be separately in equilibrium, and hence the pressure on the base equals the weight of the prism.

Since α is very small the pressure upon it may be considered uniform throughout its area, because the error introduced by this will not be comparable with the whole pressure; it is therefore at every point approximately equal to p, the value which it has at P^*: hence the whole pressure on the base is approximately $p\alpha$: also the volume of the prism is very nearly αz, and if the density of the fluid be ρ, the mass of the prism is equally nearly $\rho\alpha z$, and its weight $\rho\alpha zg$; we get therefore from the foregoing considerations,

$$p\alpha = \rho\alpha zg\ldots\ldots\ldots(1)$$

the more nearly as α is diminished indefinitely,

or $p = g\rho z$, accurately;

hence for the same fluid $p \propto z$.

NOTE. It must be remembered that the unit of force in terms of which p is here expressed is a force equal to the weight of $\dfrac{1^{\text{th}}}{g}$ of the unit of volume of the substance to which ρ refers. This is introduced in equation (1) where the weight of the prism of fluid is put equal to $g\rho \times$ its volume. It is

---

* It appears from (Art. 14) that $p\alpha$ *strictly* gives the pressure on the horizontal area α for *any* value of α; but in the proof of the proposition of this Article, we of course are not at liberty to assume the result of a subsequent proposition. Moreover, as we know nothing as yet of the form of the surface of the fluid, we must consider α indefinitely small, in order confidently to put αz as the volume of the prism.

necessary also to observe that the unit of volume is the cube, of which each edge equals the unit of length.

13*. If the upper surface of the fluid be subject to a pressure of which the measure is p_1, (1) becomes

$$p\alpha = \rho \alpha z g + p_1 \alpha, \text{ ultimately};$$

$$\therefore p = p_1 + g\rho z,$$

which determines the pressure at every point.

14. *Within a uniform fluid, which is at rest under the action of gravity alone, the pressure at every point in the same horizontal plane is the same.*

Let P, Q be *any* two points lying in the same horizontal plane below the surface of the proposed fluid; suppose a prism of fluid having PQ for its axis, and a very small transverse section α to become solid; this can in no way affect the conditions of equilibrium of the fluid. (Art. 5.)

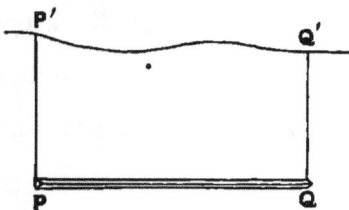

This prism is kept at rest by

the pressures on its two ends P, Q normal to these terminal planes, and therefore in the direction of PQ, *i. e.* horizontal;

the pressures upon its sides normal to these sides, and therefore perpendicular to the axis PQ;

and its own weight acting vertically downwards, and therefore also perpendicular to PQ;

therefore resolving along PQ the pressures upon the two ends, which are the only forces in this direction, must counteract each other. Let p and q be the pressures at the points P and Q, then, since α is very small, the pressure over each end of the prism will be very nearly uniform and of the same intensity as at the middle point, and therefore the pressure at the end P is $p\alpha$, and that at Q is $q\alpha$; but these must be equal, therefore

$$p\alpha = q\alpha \text{ ultimately,}$$

$$\text{or } p = q.$$

Hence, as P and Q are *any* points in the same horizontal plane, it follows that the pressure at all points in the same horizontal plane is constant.

In the preceding proposition it was assumed that PM, and in this one that PQ, lay entirely within the fluid; *i. e.* that the rigid sides of the vessel or material containing the fluid were never inclined from the vertical *towards* the body of the fluid: but these propositions are also true whatever be the form of the sides, provided the different parts of the fluid contained by them are in free communication with each other.

For let the annexed figure represent a quantity of fluid contained by the irregular sides $ABCDEF$. The circumstances of the different particles of this fluid cannot be different from what they would have been had AE merely formed a portion of a larger quantity $FAKLE$, whose surface coincides with

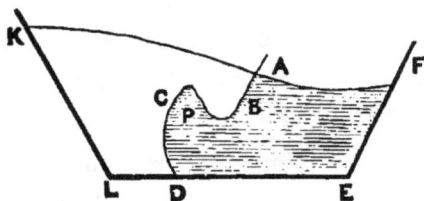

FA as far as it goes; but in this case the proof of the foregoing propositions would have held for any points in the portion $ABCDEF$: the propositions are therefore always true, P's depth in the first one being its vertical distance below the surface or the surface produced.

15. *The surface of a uniform fluid which is at rest under the action of gravity alone is horizontal.*

For taking the first figure of Article (14), if P' and Q' be the points where vertical lines through P and Q meet the surface, then by (13) $p:q::PP':QQ'$; but by Art. (14) and Cor. Art. (7) $p=q$, $\therefore PP'=QQ'$. But P and Q are any two points in the same horizontal plane below the surface of the fluid, hence any two points in the same horizontal plane within the fluid are at the same vertical distance from the surface; therefore the surface being parallel to a horizontal plane is itself horizontal.*

16. *The common surface of any two fluids at rest in the same vessel will be a horizontal plane.*

To shew this, let KL be the common surface of any two fluids in contact, AB any given horizontal plane in the one, $A'B'$ in the other: take P any point whatever in the plane AB, draw PP' vertical to meet $A'B'$ in P', cutting KL in Q.

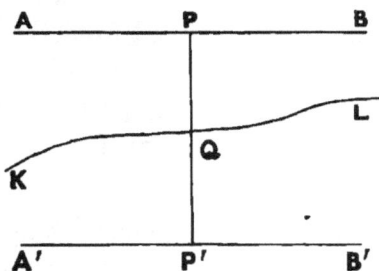

Then by Art. (14) the pressure at P will be constant for all its positions in the plane AB, call this pressure p. Similarly the pressure at P' will be constant and may be represented by p'.

Now consider the equilibrium of a prism of fluid whose axis is PP' and whose transverse section is very small and equal α; this prism of fluid is kept at rest by the normal and therefore horizontal pressures of the fluids upon its sides, the pressure downwards at P, equal to $p\alpha$, the pressure upwards at P' equal to $p'\alpha$, and its own weight;

---

* This result was not assumed in the figures of Art. (14).

∴ as these two systems, horizontal and vertical, must separately balance,

$$(p' - p)\,\alpha = \text{weight of prism};$$

call PQ, h and QP', h', and PP', a, and let the densities of the fluids, in which AB and $A'B'$ respectively are, be ρ and ρ', then since the weight of the column is the sum of the weights of the two parts PQ, and QP', if α be taken small enough, this will become $g\rho h\alpha + g\rho'h'\alpha$, hence we get

$$g\,(\rho h + \rho'h') = p' - p;$$

also $\qquad h + h' = a,$

∴ $\qquad g\,(\rho - \rho')\,h = p' - p - g\rho'a$

or $\qquad h = \dfrac{p' - p - g\rho'a}{g\,(\rho - \rho')} = \text{a constant},$

∴ the vertical distance of every point in the common surface from a given horizontal plane is the same, or the common surface is horizontal.

17. *If two heavy fluids, each of uniform density, be placed, one in each of two tubes or vessels which communicate with each other, the heights of their upper or free surfaces above their common horizontal plane of contact will be inversely as their densities.*

Suppose h and h' to be these heights, ρ and ρ' the densities of the corresponding fluids: now the pressure at any point at a depth h below the surface of the first fluid is $g\rho h$, (Art. 13); similarly the pressure due to a depth h' below the surface of the second is $g\rho'h'$; but either of these estimates must give us the pressure at a point in the plane of contact, hence they must be the same, (Art. 14);

$$\therefore \rho h = \rho'h', \text{ or } \frac{h}{h'} = \frac{\rho'}{\rho},$$

which proves the proposition.

18. *If a surface be immersed in a fluid which is kept in equilibrium by the action of gravity alone, the total normal pressure upon it is the same as would be exerted upon a plane surface of equal area placed horizontally in the same fluid at the depth of the center of gravity of the immersed surface.*

Let the area of the given surface be A, and suppose this so divided into n portions represented by $\alpha_1 \alpha_2 \ldots \alpha_n$, that by increasing n indefinitely these may all be diminished indefinitely, and may be ultimately considered plain areas all points of any one of which are at the same depth below the surface; hence the normal pressure over any area would be approximately uniform, and equal to that due to its depth: if therefore z_r be the depth of the center of gravity of the area α_r, ρ the density of the fluid, then as n is indefinitely increased, and therefore α_r diminished, the normal pressure upon α_r continually approaches to $g\rho z_r \alpha_r$.

The same thing will be true for each of the other portions into which the surface has been divided; hence the total normal pressure will upon the same supposition approach the sum of the terms

$$g\rho z_1 \alpha_1 + g\rho z_2 \alpha_2 + \ldots + g\rho z_n \alpha_n,$$

or if we represent this pressure by P, then

$$P = g\rho \left(z_1 \alpha_1 + z_2 \alpha_2 + \ldots + z_n \alpha_n \right) \ldots \ldots \ldots (1) \text{ ultimately.}$$

But if $\bar{z}$ be the depth of the center of gravity of the whole area A below the surface, then

$$\bar{z}A = z_1 \alpha_1 + z_2 \alpha_2 + \ldots + z_n \alpha_n \ldots \ldots \ldots (2).$$

This is true whatever be the magnitudes of α_1, α_2, &c.; it will therefore be true when they are indefinitely small, which was the condition by which the equation (1) was obtained, and we may substitute from (2) into (1): we thus get

$$P = g\rho \bar{z} A;$$

but the right-hand side of this equation is evidently the pressure which would be exerted upon a surface A immersed at a uniform depth $\bar{z}$ in the fluid in question, and hence the truth of the proposition.

This proposition admits of the following statement:

If a surface be immersed in a fluid which is kept in equilibrium by the action of gravity alone, the total normal pressure upon it is equal to the weight of a prism of the same fluid, of transverse section equal to the area of the given surface and of altitude equal to the depth of the center of gravity of the immersed surface.

It appears that P is given in terms of the unit of weight of Art (10).

18*.   If the surface of the fluid be subject to a uniform pressure of which p_1 is the measure, the total pressure on the immersed surface A will be

$$p_1 A + g\rho\bar{z}A.$$

This will be understood at once on considering that the pressure on the surface is transmitted with equal intensity in all directions through the fluid. (Art. 7.)

19.   The last proposition gives only the *sum* of the normal pressures upon a surface of a body immersed in a fluid which is acted upon by gravity alone: *the resultant of the same pressures is equal to the weight of the fluid displaced by the body, its direction is vertical, and passes through the center of gravity of the fluid so displaced.*

Let Q represent the portion of the body which is immersed, whether it be totally so or not. It is evident that the pressures upon the surface of Q in contact with the fluid depend only upon the position and extent of that surface,

and not at all upon the nature of Q itself: they will therefore be unaltered if any other
body conterminous with Q be
substituted for it. Suppose then
some of the fluid under consideration to be solidified into the
form of Q and placed in its
stead, the pressures on the surface of this solidified fluid are
the same as those on Q; but this solidified fluid so placed
will be at rest, for it would be so if it were not solid, and
its solidification cannot affect equilibrium, (Art. 5): now the
only forces acting upon this solidified fluid are its own
weight, vertically downwards through its center of gravity,
and the before-mentioned fluid pressures upon its surface,
hence the resultant of these pressures must be equal and
opposite to this weight. The fluid which we have supposed
solidified is that which would exactly fill the place of the immersed portion of the body, if the body were removed; it is
generally spoken of as the fluid displaced by it: the proposition is therefore proved.

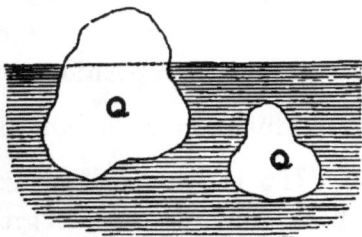

20.   *When a body floats in a fluid under the action of
gravity only, the weight of the fluid displaced by it is equal to
its own weight, and the centers of gravity of the fluid displaced
and of the body itself are in the same vertical line.*

The only forces acting upon the body are its own weight,
in a vertical direction at its center of gravity, and the pressures
of the fluid upon the surface immersed: hence since there is
equilibrium, the resultant of these pressures must be equal
and opposite to the weight of the body, and it must act
vertically upwards through the center of gravity of the body;
but by the last proposition the resultant of these pressures
was shewn to be equal to the weight of the fluid displaced

and to act vertically upwards through the center of gravity of the fluid displaced. From these two assertions, then, we obtain that: the weight of the fluid displaced equals the weight of the floating body, and that the centers of gravity of the two lie in the same vertical line.

21. If a floating body be disturbed in its position in such a way that the amount of fluid displaced by it remains the same, the forces acting upon it will be unaltered as regards magnitude and direction, for they will be, its own weight acting vertically downwards at its center of gravity and the weight of the displaced fluid, which is, as before, equal to the weight of the body, and acts vertically upwards through its center of gravity: but in the general case these two centers of gravity will be no longer in the same vertical line, and thus a couple will have been produced, under whose action the body will either return to its original position of equilibrium or will be removed further from it, according as the new direction of the resultant of the fluid pressures, i. e. the weight of the fluid displaced, meets that fixed line in the body, which passes through its center of gravity and was vertical in the body's floating position, above or below the center of gravity. This

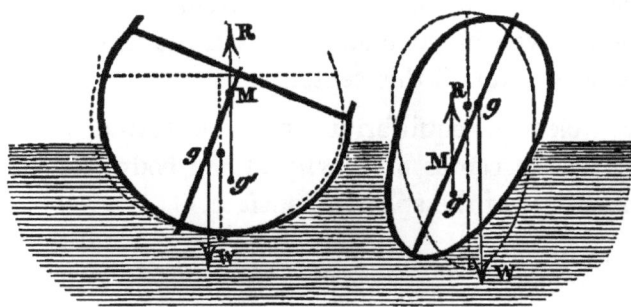

is made evident by the annexed figure, where the body is represented in its disturbed position, g is its center of gravity,

g' that of the fluid displaced, W the weight of the body, R the resultant of the fluid pressures on its surface and therefore equal W, M the point where the direction of R meets the fixed line through g,—the dotted figures refer to the original position of equilibrium.

The original position of the floating body is said to be one of *stable* equilibrium, when upon a *very slight* disturbance of this kind the couple produced tends to bring the body back again, and *unstable* when the contrary is the case: instances of these two are given in the figure. The equilibrium is said to be neutral whenever this very small displacement fails to produce a couple, i. e. when the two centers of gravity are still brought by it into the same vertical line.

It is not difficult to see that when a body floats with its center of gravity below that of the fluid displaced, the equilibrium will be stable.

Ex. In illustration of this article, consider the case of a floating body, of which the portion immersed is part of a sphere.

The direction of the fluid pressures being normal will at every point pass through the center of this spherical surface.

Therefore the direction of the resultant of the fluid pressures must, as well in the disturbed as in the floating position, pass vertically through this center.

Hence, clearly, equilibrium will be stable or unstable according as the center of gravity of the body falls below or above the center of the spherical surface.

22. *If a body be immersed in a fluid of less specific gravity than itself it will sink.*

Let V be the volume of such a body Q; S, S' the specific gravities of the fluid and body respectively; then the forces acting upon Q are its own weight $= VS'$ acting vertically downwards through its center of gravity, and the resultant of the fluid pressures upon its surface; but this resultant is equal to the weight of the fluid displaced by Q, $= VS$, and acts vertically upwards through the center of gravity of the fluid displaced, which is also that of the body, if we suppose the body and fluid each to be of

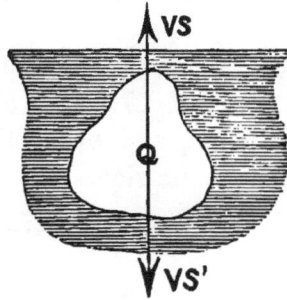

uniform density; therefore on the whole the body is acted upon by a vertical force equal to the difference between these two, i. e. of the weight of the body and that of the fluid displaced by it, $= V(S' - S)$ in a downward direction; it must therefore sink.

23. If it be required to find the force to be applied by means of a string in order to hold the body in its position, it must evidently be equal and opposite to this force $V(S' - S)$.

But the force requisite to support a heavy body or to keep it from falling under the action of gravity is taken as the measure of its weight. Hence the foregoing shews that *the apparent weight of a body when immersed in a fluid is less than its real weight by the weight of the fluid displaced.* This result is very useful in finding the specific gravities of bodies.

24. If on the contrary the specific gravity of the body immersed be less than that of the fluid, it will rise: for, as before, the resultant force upon the body is a single vertical force passing through its center of gravity, equal to the difference between the weight of the body and that of the fluid displaced by it, and acting in the direction of the larger

force, which in this case is upwards.   This will serve to
explain the ascent of a balloon.

25.   If the specific gravities of the body and fluid be the
same, i. e. if S and S' be equal, this resultant force clearly
vanishes, and hence the body would rest in any position of
total immersion.

The remainder of this section gives some methods of
comparing the specific gravities of different substances
whether solid or fluid, and describes instruments called
Hydrometers, which are used for the purpose: it may be
remarked that in all cases the ratio between the weights
of equal volumes of the two substances is the quantity
sought. (Art. 11.)

26.   An ordinary balance adapted to weighing bodies in
fluids is sometimes termed an Hydrostatic balance: one of the

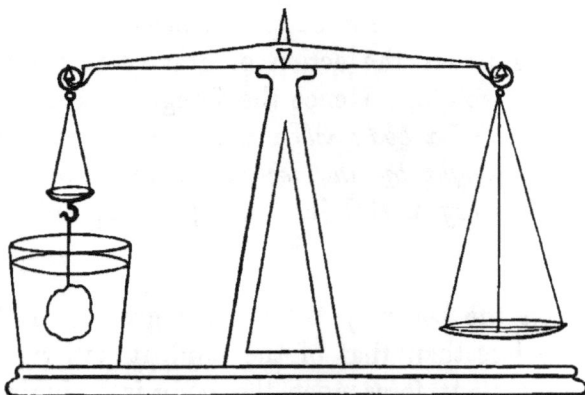

scales is small and hung very short; at the bottom of the
pan is a hook from which the body, while immersed in a fluid

contained in any vessel below it, may be suspended by a small string or wire.

27. *To find the specific gravity of a solid body, that of distilled water being taken as the unit.*

(1) When the specific gravity of the body is greater than that of the distilled water, or in other words, when the body sinks upon immersion :

Let the weight of the body in vacuum be determined $= W$ suppose; then let its apparent weight in distilled water be ascertained by the hydrostatic balance, suppose it $= W'$; then $W - W'$ is (Art. 23) the weight of the distilled water displaced by it; if then S be the specific gravity required, (Art. 11),

$$S = \frac{\text{weight of the body}}{\text{weight of an equal volume of distilled water}}$$

$$= \frac{W}{W - W'}.$$

(2) When the specific gravity of the body is less than that of the distilled water :

Let a piece of some heavy substance be attached to the body, such that the whole will sink upon immersion; let w be the ascertained weight of this attached portion in vacuum, w' in the water, W_1 the weight of the compound body in vacuum, W_1' the weight of the same in the water, and W, as before, the weight of the body itself in vacuum; then

the weight of water displaced by the compound body when immersed $= W_1 - W_1'$

of that displaced by the attached body $= w - w'$;

hence the weight of water displaced by the proposed body, which must be the difference between these, is

$$(W_1 - W_1') - (w - w');$$

therefore by the preceding case

$$S = \frac{W}{W_1 - W_1' - w + w'}.$$

28. *To find the specific gravity of a fluid.*

Let a vessel be filled with it and then weighed in vacuum, and let the weight so found be W; let also the weight of the vessel itself when quite empty be W', and when filled with distilled water W'': if care be taken to fill the vessel accurately, the volumes of the proposed fluid and of the water weighed will be the same; now the weight of the first is $W - W'$, and of the latter $W'' - W'$, hence if S be the specific gravity required,

$$S = \frac{W - W'}{W'' - W'}.$$

By this method also the specific gravities of air, or any gases, or even of very fine powders, may be obtained.

29. *The specific gravities of two fluids may be compared by weighing the same solid in each.*

Thus, let W be the weight of the solid in vacuum,

W_1 its weight when immersed in the first fluid,

W_2 .... when .... in the second fluid,

then $W - W_1$ is the weight of the first fluid displaced by it,

$W - W_2$ is the weight of the second fluid displaced by it;

but these are the same in volume, therefore if S_1, S_2 be their specific gravities,

$$\frac{S_1}{S_2} = \frac{W - W_1}{W - W_2}.$$

30. Of the weights which enter the preceding formulæ, those stated to be found by weighing in a vacuum, are very properly termed *true* weights; those found by weighing in any medium are usually called *apparent* weights. The difference between the true and apparent weights is affected by two causes, which tend to counteract each other; on the one side, the weight which the body requires to balance it is less than what it would require in vacuum by the weight of the medium which it displaces; on the other, the balancing body is less than the weight which it is supposed to represent by the weight of the medium which itself displaces; thus if a *true* 1 lb. weight and a body Q when placed in the two scales of a balance in the air keep the beam horizontal, it can only be concluded that the 1 lb. weight diminished by the weight of air which it displaces, is equal to the weight of Q diminished by the weight of air which Q displaces; it will be seen that when the weight of air displaced by the two bodies is the same, the weight of Q is 1 lb. and not otherwise.

However, since the specific gravity of air is not greater than $\frac{1}{800}$, water being the standard substance, therefore the apparent weight in air and the true weight of a substance whose specific gravity is not small, will differ *very slightly*, and may practically be considered equal.

31. In the foregoing methods of finding the specific gravities of substances we have deduced them by considering them to be proportional to the weights of equal volumes; it may be seen from the formula (3) of Art. (11), that they are also inversely proportional to the volumes which have equal

weights. The two hydrometers which are most generally employed, are constructed respectively upon these two principles.

The Common Hydrometer

gives the ratio between the volumes of two liquids which have the same weight: the annexed figure represents it. AB is a thin graduated stem of uniform transverse section, which at its lower extremity expands into a hollow globe BC; and to this is fixed a ball of lead D sufficiently large to bring the center of gravity of the whole instrument within it.

To apply the instrument, it is immersed in a proposed fluid and allowed to find its position of equilibrium, which it will very readily do on account of the lowness of its center of gravity, (Art. 19); the stem will be vertical, and the number of its graduations cut off by the surface of the fluid can be easily observed; the comparison of this number with that given by immersion in the other fluid will lead us to the ratio between their specific gravities; for let S and S' be these, W the whole weight of the instrument, V its volume, and a_1 the transverse area of the stem AB: when the instrument is immersed in the first fluid snppose it to sink to P, then

$$W = S(V - a_1 AP);$$

again, when immersed in the second fluid suppose it to sink to P'; then

$$W = S'(V - a_1 AP');$$

$$\therefore \frac{S}{S'} = \frac{V - a_1 AP'}{V - a_1 AP}:$$

hence the ratio between S and S' is known when the numbers of graduations in AP and AP' are known.

Nicholson's Hydrometer.

32. This instrument is adapted for finding the ratio between the weights of equal volumes of two fluids, or between the weights of a solid and an equal volume of fluid. It is represented in the annexed figure; BC is a hollow buoyant body of any symmetrical shape, A is a cup supported upon it by a rigid wire AB, and D is a similar cup suspended below by a wire CD. This cup is frequently capable of inversion, so as to hold down a body specifically lighter than the fluid.

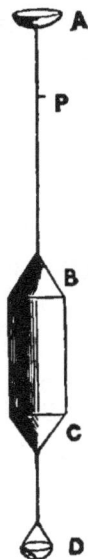

(1) To use this instrument for comparing the specific gravities of two fluids.

Let W be the weight of the instrument, W_1 the weight which must be placed in A in order to make it sink in the first fluid to a point P in the stem AB, W_2 the weight to be placed in A in order to make it sink to the same point in the second fluid: then the weight of the fluid displaced in the first case is $W + W_1$, in the second $W + W_2$, and the volumes are the same in both; therefore if S and S' be the specific gravities whose ratio is required,

$$\frac{S}{S'} = \frac{W + W_1}{W + W_2}.$$

(2) To compare the specific gravities of a solid and fluid:

Let W_1 be the weight required to be put in A in order to sink the instrument up to P in the fluid; remove this and place the solid in A; and let W_2 be the weight which must in addition be put in A in order to sink the hydrometer in the fluid to the same point P; next place the solid in D, and let

W_3 be the weight which must now be put in A to sink the instrument to the same point;

then the weight of the solid is $W_1 - W_2$,

also its apparent weight in the fluid is $W_1 - W_3$;

but this must be its real weight diminished by the weight of the fluid it displaces; ∴ the weight of fluid displaced by it is $W_1 - W_2 - (W_1 - W_3)$

$$= W_3 - W_2:$$

since then this displaced fluid is equal in volume to the displacing solid, if S and S' represent the respective specific gravities required,

$$\frac{S}{S'} = \frac{W_1 - W_2}{W_3 - W_2}.$$

33.　When two or more fluids are thrown together in the same vessel, if they do not lie in masses superimposed so that the common surfaces are horizontal planes (Art. 16) they will become so intimately mixed as to form a new fluid.

In this case, if the fluids be incompressible the specific gravity of the compound will be known when that of each of the composing fluids is so. For let $V_1, V_2, ... V_n$ be the volumes of the different fluids thus mixed together, $S_1, S_2, ... S_n$ their respective specific gravities; $V_1 + V_2 + ... + V_n$ is the volume of the resulting mixture, and if S be its specific gravity, since the weight of the whole must equal the sum of the weights of the parts,

$$(V_1 + V_2 + ... + V_n)\, S = V_1 S_1 + V_2 S_2 + ... + V_n S_n,$$

and therefore

$$S = \frac{V_1 S_1 + V_2 S_2 + ... + V_n S_n}{V_1 + V_2 + ...+ V_n}.$$

EXAMPLES TO SECTION II.

(1) Let ABC be a rigid pipe of small bore, communicating at C with a vessel DCE, whose top DE is moveable up and down by some means which allows of the vessel remaining watertight: it may be a piston fitting closely to a cylinder, or it may be more simply a board connected with the bottom by leather sides. If the whole of this be filled with water, it is found that a man may easily support himself upon DE, by merely closing the top of AB with his finger, or he may even raise himself by blowing into AB from his mouth. This phenomenon is sometimes called the *Hydrostatic Paradox :* explain it.

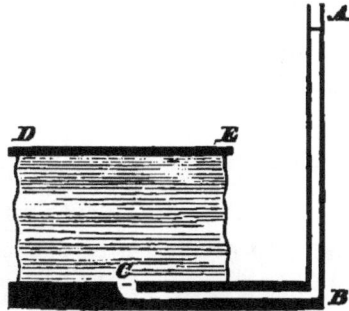

When the man applies his finger to A, he presses the surface of the water in the pipe with a certain force, which (by Art. 7) is thence transmitted through the fluid to every portion of surface in contact with it: if then α be the cross section of the tube at A and P the force he applies, a force equal to P will be transmitted to every portion of DE which is equal to α; but if A be the whole area of DE, it contains $\dfrac{A}{\alpha}$ such portions; therefore the whole force applied upwards to $DE = \dfrac{A}{\alpha} P$, which may be quite large enough to support the man's weight, although P is small, provided the area α be small compared with A, and therefore the fraction $\dfrac{A}{\alpha}$ a very large number.

If P be increased beyond this previously supposed value, by blowing or otherwise, the man will evidently raise himself.

Ex.   Find P that it may just support the man's weight W.

(2)   The pressure at a point P within a body of water, under the action of gravity only, is 50 lbs.; given that the weight of a cubic foot of water is 1000 oz., and that the unit of area is a square foot, find the depth of P below the surface.

Let z be this depth in feet, then (by Art. 13) if ρ be the density of the water and p the pressure at P,

$$p = g\rho z \times \frac{\text{weight of unit of volume of standard substance}}{g};$$

$\therefore$ by the question, considering a foot as the unit of length and water as the standard substance, and $\therefore$ $\rho = 1$,

$$50 \text{ lbs.} = z \times \frac{1000}{16} \text{ lbs.}$$

$$\therefore z = \frac{4}{5},$$

or the required depth is $\frac{4}{5}^{\text{ths}}$ of a foot.

(3)   A cylinder is immersed in water in such a way that its axis is vertical and its top is just level with the surface; find the total normal pressure upon its bottom and sides.

By Art. (16), this total pressure is equal to the weight of a cylindrical column of water whose base equals the area of the given surface pressed, and whose height is equal to the depth of the center of gravity of this given surface below the surface of the water.

But if r be the radius of the base and h the height of our cylinder, the area of the surface pressed is

$$= \text{area of base} + \text{area of sides}$$

$$= \pi r^2 + 2\pi rh.$$

Again, the depth of the center of gravity of this pressed surface below the top of the water

$$= \frac{\pi r^2 . h + 2\pi rh . \dfrac{h}{2}}{\pi r^2 + 2\pi rh};$$

∴ the column of fluid whose weight is sought has a volume

$$= (\pi r^2 + 2\pi rh) \times \frac{\left(\pi r^2 h + 2\pi r \dfrac{h^2}{2}\right)}{\pi r^2 + 2\pi rh}$$

$$= \pi r^2 h + 2\pi r \frac{h^2}{2};$$

∴ the pressure required is $g\rho \left(\pi r^2 h + 2\pi r \dfrac{h^2}{2}\right)$

$$= \pi g\rho rh (r + h),$$

ρ being the density of the water.

(4) A triangle ABC is immersed in a fluid, its plane being vertical and the side AB in the surface: if O be the center of the circumscribing circle, prove that the pressure on the triangle OCA : pressure on triangle OCB :: $\sin 2B$: $\sin 2A$.

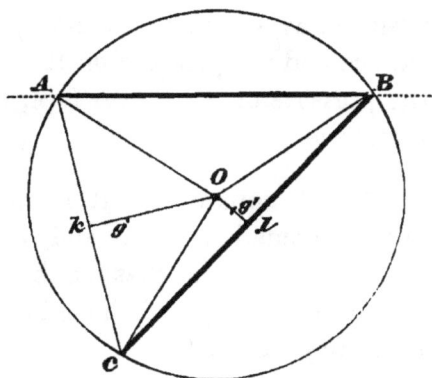

The pressures on the two triangles will be to each other in the same proportion as the product of the area of each triangle into the depth of its center of gravity below AB, (Art. 18).

But if g and g' be these centers of gravity, they will divide the lines Ok, Ol, drawn from O to the points of bisection of AC and BC respectively, in the same proportion; therefore they will be in a straight line parallel to that joining k and l, and therefore parallel to AB.

Hence the pressures required will be as the areas only, i. e. pressure on OCA : pressure on OCB.

$$:: \text{area } OCA : \text{area } OCB$$
$$:: \sin AOC : \sin BOC$$
$$:: \sin 2ABC : \sin 2BAC.$$

Q. E. D.

(5) A regular hexagon is immersed vertically in a fluid, so that one side coincides with the surface; compare the pressures on the triangles into which it is divided by lines drawn from its center to the angular points.

(6) A cylinder whose height is 4 feet is sunk in water with the axis vertical till its upper face is 805 feet below the surface and the pressure on the top is found to be 35 lbs.; find the pressure on the lower face, neglecting the pressure of the atmosphere.

(7) A square is just immersed in a fluid of density 8, with one side horizontal and with its plane inclined at 60° to the vertical: given that a cubic foot of the standard substance weighs 1000 ozs., find the side of the square that the pressure on it may be 216 lbs.

(8) A vessel containing water is placed on a table; supposing the vessel of such a shape that only half the fluid is vertically over its base, what is the pressure on the base? Is this the pressure on the table? Explain your answer.

The reasoning of Art. (4) aided by a reference to the second figure of Art. (14) will explain how a rigid surface may supply the place of a vertical column of fluid. The rigidity is the result of internal forces, and does not affect the pressure on the table.

(9) The same quantity of fluid which will just fill a hollow cone is poured into a cylinder whose base equals that of the cone: compare the pressures on the bases, the axes of both vessels being vertical.

If the cone and cylinder be resting on a horizontal plane, state how the pressures on this plane will be affected, and explain the case fully.

(10) Suppose a pound weight of a substance twice as specifically heavy as water to be hung into the water contained in a vessel, which is standing on a table, by a string not attached to the vessel, what would be the increase of pressure on the table?

(11) A cylinder of given radius, height, and specific gravity, is partially immersed with its axis vertical in water, being held up by a string which is attached by one end to its top, and by the other to a fixed point vertically above the cylinder: supposing the string to stretch 1 inch for every 5 lbs. which it supports, and that its unstretched length just allows the bottom of the cylinder to touch the water, and that a cubic foot of water weighs 1000 ozs., find the depth of immersion.

Let this depth be z feet: also let h be the height of the cylinder and r the radius of its base in feet, σ its specific gravity.

Then the volume of water displaced by the cylinder is $\pi r^2 z$ cubic feet, and therefore the weight of it, which is the same as the resultant of the fluid pressures upwards upon the cylinder, must be (Art. 19)

$$= \pi r^2 z \frac{1000}{16} \text{lbs.}$$

Also, since the string is stretched z feet, its tension must by question

$$= 12z \times 5 \text{ lbs.}$$

Now these two forces, each acting vertically upwards upon the cylinder through its center of gravity, and the weight of the cylinder itself acting vertically downwards through the same point, are the only forces which are acting upon the cylinder; therefore, for equilibrium it is only necessary that the sum of the first two equal the last; but the weight of the cylinder $= \pi r^2 h \times \sigma \times \dfrac{1000}{16} \text{lbs.}$,

$$\therefore z \left(\pi r^2 \frac{1000}{16} + 60 \right) = \pi \sigma r^2 h \times \frac{1000}{16} ;$$

$$\text{or } z = \frac{\pi \sigma r^2 h}{\pi r^2 + \dfrac{24}{25}} \text{ feet.}$$

It should be observed that all the symbols here used are necessarily by the statement *numerical* quantities.

(12) What weight is just sufficient to hold down a balloon containing 2500 cubic feet of hydrogen gas (specific gravity .069 referred to air) supposing the weight of the

material enclosing the gas 4 lbs. and the weight of a cubic foot of common air 1.1 oz.?

(13) A cylinder which floats in water under an exhausted receiver has $\frac{3}{4}$ of its axis immersed; find the alteration in the depth of immersion when air, whose specific gravity is .0013, is admitted.

(14) A cone 7 inches in height and 2 inches in diameter at its base is attached to a hemisphere of equal diameter: the specific gravity of the cone is 1.5, that of the hemisphere is 1.75; find the specific gravity of the fluid in which this compound body will sink to a depth of 3 inches with the vertex of the cone upwards.

(15) When 30 ozs. of an acid A whose specific gravity is 1.5 are mixed with 35 ozs. of an acid B whose specific gravity is 1.25, and with 35 ozs. of water, the specific gravity of the resulting mixture is found to be 1.35; find the contraction of volume, assuming the specific gravity of water to be 1, and the weight of a cubic foot of its volume to be 1000 ozs.

The volume of 35 ozs. of water $= \dfrac{35}{1000}$ cubic feet,

$$30 \ldots\ldots\ldots\ldots A = \dfrac{30}{1.5 \times 1000} \ldots\ldots;$$

$$35 \ldots\ldots\ldots\ldots B = \dfrac{35}{1.25 \times 1000} \ldots\ldots$$

Also the volume of the 100 ozs. of mixture

$$= \dfrac{100}{1.35 \times 1000} \text{ cubic feet};$$

∴ the loss of volume is

$$\left(35 + \frac{30}{1.5} + \frac{35}{1.25} - \frac{100}{1.35}\right)\frac{1}{1000} \text{ cubic feet.}$$

(16) A man whose weight is 168 lbs. can just float in water when a certain quantity of cork is attached to him. Given that his specific gravity is 1.12, that of cork .24, and that of water 1, find the quantity of cork in cubic feet, assuming a cubic foot of water to weigh 1000 ozs.

Let V be the required number of cubic feet of cork, then $V \times .24 \times \frac{1000}{16} + 168$ is the weight of the man and cork together in lbs.

By the question this must be just equal to the weight of the same volume of water; and the volume of the man in cubic feet is $\frac{168 \times 16}{1.12 \times 1000}$, because each cubic foot of him weighs 1.12×1000 ozs. (Art. 11), and his whole weight is given to be 168 lbs. ;

$$\therefore \left(V + \frac{168 \times 16}{1.12 \times 1000}\right)\frac{1000}{16} \text{ lbs.} = \left(V \times .24 \times \frac{1000}{16} + 168\right) \text{ lbs.,}$$

$$V(1 - .24)\frac{1000}{16} = 168 - \frac{168}{1.12}$$

$$= \frac{12}{112} \times 168,$$

$$V = \frac{16 \times 12 \times 168}{112 \times .76 \times 1000} = \frac{16 \times 12 \times 168}{112 \times 760}$$

$$= \frac{6 \times 168}{7 \times 380} = \frac{6 \times 42}{7 \times 95};$$

$$\therefore V = .37895 \text{ of a cubic foot.}$$

(17) If in a circular tube two fluids be placed so as to occupy 90° each, and if the diameter joining the two open surfaces be inclined at 60° to the vertical, prove that the densities are as

$$\sqrt{3} + 1 : \sqrt{3} - 1.$$

Let $ACBD$ be the tube; AD, BD the portions of it occupied by the two fluids whose specific gravities may be represented by ρ and ρ' respectively: if the diameter AOB be drawn, it will be inclined to the vertical at an angle 60°. Let the horizontal line through D, the common surface of the two fluids, meet the verticals through A, O, and B respectively in M, N, and P, and draw Af, Bg perpendicular to ON.

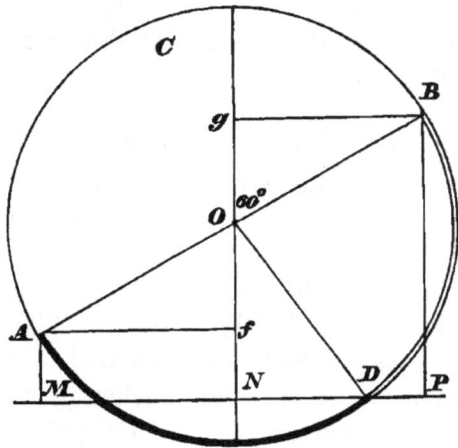

Then, by Art. (33),

$$\frac{\rho}{\rho'} = \frac{BP}{AM} = \frac{ON + Og}{ON - Of} = \frac{OD \sin 60° + OB \cos 60°}{OD \sin 60° - OA \cos 60°}$$

$$= \frac{\tan 60° + 1}{\tan 60° - 1}$$

$$= \frac{\sqrt{3} + 1}{\sqrt{3} - 1}.$$

(18) Equal volumes of oil and alcohol are poured into a circular tube so as to fill half the circle, shew that the common surface rests at a point whose angular distance from the lower point is $\tan^{-1}\frac{12}{171}$; the specific gravities of oil and alcohol being .915 and .795.

(19) A body P weighs 10 lbs. in air and 7 lbs. in a fluid A : if it be attached to a denser body Q, and then suspended in another fluid B, the apparent weight of both bodies is 5 lbs. less than that of Q alone; compare the specific gravities of A and B.

(20) A cone floats in a fluid with its axis vertical, the vertex being downwards and half its axis immersed; compare the specific gravity of the cone with that of the fluid.

(21) The specific gravities of sea-water, olive-oil, and alcohol are 1.027, .915, and .795; the oil and alcohol have depths one inch and two inches above the water. Find the pressure on 3 square inches of a plane surface which is immersed horizontally at a depth of 5 inches below the upper surface of the oil: the weight of a cubic foot of distilled water being 1000 ozs.

(22) If s be the specific gravity referred to water of a body whose bulk is n cubic inches and weight m ozs., then

$$m \times 1728 = 1000 \times n \times s.$$

(23) The mean specific gravity of a plated cup is 7.6; that of the silver is 10.45; that of the unplated metal 7.3; compare the volumes and weight of the metals.

(24) The specific gravity of zinc is 6.862, what is the weight of the water displaced by a portion of it, which when immersed weighs 5.862 lbs.?

(25) The volume between two successive divisions of the stem of a hydrometer is $\frac{1}{1000}$th part of the bulk of the whole instrument; it floats in water with 20 divisions above the surface: find the least specific gravity of a fluid in which it will float.

(26) A hydrometer that weighs 250 grains, requires 94 grains to sink it in water to the requisite point, and 8 grains in naphtha; when a substance is placed successively in the upper and lower cup, $1\frac{1}{2}$ grains and 14 grains are respectively sufficient to sink the instrument in naphtha to the requisite point; required the specific gravity of the substance.

(27) How many inches are there in the edge of a cubical mass of coal which weighs 2 tons, its specific gravity being 1.12? and what is the specific gravity of silver one cubic inch of which weighs 6.1 ozs.? also what is the weight in ozs. of 30 cubic inches of mercury, its specific gravity being 13.6?

(28) If $w_1 w_2 w_3$ be the apparent weights of a body when weighed in three fluids whose densities are respectively $\rho_1 \rho_2 \rho_3$, shew that

$$w_1 (\rho_3 - \rho_2) + w_2 (\rho_1 - \rho_3) + w_3 (\rho_2 - \rho_1) = 0.$$

(29) Two metals of which the specific gravities are 11.22 and 7.25, when mixed in certain proportions without condensation, form an alloy whose specific gravity is 8.72; find the proportion by volume of the metals in the alloy.

(30) A small vessel when *entirely* filled with distilled water weighs 530 grains; 26 grains of sand are thrown into the vessel, and the whole then weighs 546 grains. Shew that the specific gravity of the sand is 2.6.

(31) A crystal of saltpetre weighs 19 grains: when covered with wax (the specific gravity of which is .96) the whole weighs 43 grains in vacuo and 8 grains in water. Shew that the specific gravity of saltpetre is 1.9.

(32) 37 lbs. of tin loses 5 lbs. in water, 23 lbs. of lead loses 2 lbs. in water, a composition of lead and tin weighing 120 lbs. loses 14 lbs. in water; find the proportion of lead to tin in the composition.

(33) A solid hemisphere turning round a fixed horizontal axis fits into a fixed hemispherical cup: shew that if the hemisphere be turned through any angle, and the cup then filled up with fluid of double the specific gravity of the solid, the solid will rest in that position.

Let *ADB* be a section of the hemispherical cup made

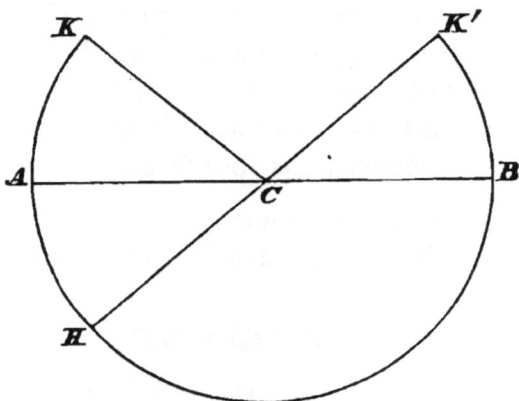

by the plane of the paper, perpendicular to the fixed axis about which the solid turns, *C* this axis, and *HDBK'* the solid turned through any angle *ACH*. If the part *AHC* of the cup be now filled up with fluid whose specific gravity is double that of the solid, equilibrium will be preserved.

For it is manifest that equilibrium would obtain, if the space *HCK* were filled up with a portion of the same substance as the solid hemisphere. Also, since the centers of gravity of the two figures, *HCA*, *ACK* are necessarily in the same vertical line, the effect of *HCK* in producing equilibrium must be the same as a uniform solid *HCA* whose weight is equal to the sum of the weights of *HCA* and *ACK* together, *i. e.* as a uniform solid *HCA* having double the specific gravity of the given solid. But since *AC* is horizontal, no new circumstances affecting the pressure on *HC* would be introduced by supposing this solid to become liquid, which would produce the proposed case: hence the truth of the proposition.

(34) Is it advantageous to a buyer of diamonds that the weighing of them should be made when the barometer is high or when it is low, supposing their specific gravity to be less than that of the substance used as the weight? (Art. 30.)

(35) A sphere of weight W with its center of gravity bisecting a radius floats in a fluid: W' is the weight of a volume of the fluid equal to the volume of the sphere; shew that if $W' > 3W$, there is a point within the sphere where a weight may be placed so that the sphere may float in any position with half its volume immersed.

(36) Find the vertical angle of an isosceles triangle in order that, when floating with an angle at the base downwards in any fluid of greater specific gravity than itself, the opposite side may be horizontal.

(37) A cylinder $(s.g.\ \sigma)$ floats with its axis vertical partly in one fluid $(s.g.\ \sigma_1)$, partly in another $(s.g.\ \sigma_2)$; shew that the common surface divides the axis in the proportion of $\sigma - \sigma_2 : \sigma_1 - \sigma$.

(38) A rod of density ρ and length a is freely moveable about one end fixed at a depth c below the surface of a fluid of density σ: shew that the rod may rest in a position inclined to the vertical provided that

$$\frac{\sigma}{\rho} > 1 < \frac{a^2}{c^2},$$

and that such a position is stable.

SECTION III.

34. ELASTIC fluids are either permanent gases, or vapours which upon a sufficient reduction of temperature assume the state of liquids: of these last steam may afford an example. There is however no hydrostatical distinction between gases and vapours, indeed there is every reason to think that they only differ physically in the range of circumstances under which they may be severally considered permanent.

As is the case in inelastic fluids, so the pressure at any point within an elastic fluid is due partly to the transmission of forces from the surfaces of the fluid, according to the law of (Art. 7), and partly to the direct action, at that point, of gravity or any other external force. But the one great distinction between the two kinds of fluids is, that in the inelastic this pressure does not alter the relative distances of the adjacent particles from each other; they appear capable of affording reactions to any required amount, as is the case with rigid surfaces in contact, without their geometrical relations to each other being affected; while in elastic fluids, the reactions between adjacent particles seem to depend upon their mutual distances, the greater the force thus required to be called forth the nearer the particles approach each other, and the smaller the volume of the mass becomes (Art. 8); upon the diminution of this force they again recede, and the volume increases. The resultant of these reactions may therefore very well be termed the *elastic force* of the fluid at that point:

an experiment directed to ascertain its relation to the state of compression of the particles, *i. e.* to the density of the fluid, will presently be described.

(35)   The weight of elastic fluids such as air, is generally, for the ordinary volumes and densities which come under our notice, so small that it may be entirely omitted in comparison with the transmitted forces; the pressure then becomes uniform for every point throughout the mass, unless the circumstances of the case introduce other external forces.   But the effect of gravity upon the mass of air contained in the enormous volume of the atmosphere produces a pressure at the earth's surface which can never be neglected; its amount may be easily estimated by the aid of

The Barometer.

36.   Suppose a tube *BA* of considerable length and filled with mercury, to be inverted into a vessel *DG* also containing mercury; and if the end *B* remain closed and *A* be opened, the mercury in the tube will be observed to sink to a certain point *C* and no farther, leaving a vacuum in the upper part *BC* of the tube; let *DKE* be the common surface of the external air, and the mercury; then the pressure at every point of this must be the same (Art. 14); that at any point without the tube is due to the weight of the atmosphere, call it Π, and at any point within the tube it is due to the weight of the column *CK* of mercury; hence if σ be the specific gravity of mercury, and h be the vertical height of *KC*, we must have (Art. 13)

$$\Pi = \sigma h.$$

Hence since σ, the specific gravity of mercury, may be generally considered constant, we get

$$\Pi \propto h,$$

or h will serve to measure Π, the pressure of the atmosphere on a unit of area.

Any instrument, as we have here described, furnished with graduations or any other means of observing the length KC or h, is called a Barometer. It is obvious that any other heavy inelastic fluid might be used instead of mercury, but as h varies inversely as σ for a given value of Π, it is always advantageous to employ as heavy a fluid as possible, because the instrument becomes very awkward when BK is required to be long; even with mercury the length of h is about 30 inches for the ordinary pressure of the atmosphere. \

Note. σ is not absolutely constant, since the volume of mercury, and therefore its specific gravity, changes with a change of temperature. This variation being very slight is neglected in rough observations. If, however, close accuracy be required the thermometer must be noted at the time, and from the observed temperature and a known law connecting temperature and volume, may be determined the height at which the column would stand for a given standard specific gravity of mercury.

In the adjustment of the Barometer it is essential that the graduated scale, by which h is measured, should be vertical.

For further observations on the Barometer, see Art. 71.

The following two examples are given to illustrate the equation

$$\Pi = \sigma h.$$

(1) Find the atmospheric pressure on the square inch, when the height of the mercurial barometer is 30 inches,

assuming the specific gravity of mercury, referred to water to be 13.6.

From the equation

$$\Pi = \sigma h$$

we have, if an inch be taken as the unit of length, the pressure on the square inch $= \sigma h$ times the weight of a cubic inch of water,

$$= 13{\cdot}6 \times 30 \times \frac{1000}{1728} \text{ ozs.}$$

since a cubic foot of water weighs 1000 ozs. nearly,
or the required pressure $= 14\frac{3}{4}$ lbs. nearly.

(2) Given that the pressure of the atmosphere on the square inch is P lbs., find the height of the barometer.

In this case

$$\Pi = \sigma h \text{ gives us}$$

P lbs. $= \sigma h$ times the weight of a cubic inch of water

$$= 13{\cdot}6 \times h \times \frac{1000}{1728} \cdot \frac{1}{16} \text{ lbs. nearly;}$$

$$\therefore \ h = \frac{864}{425} \cdot P \text{ inches.}$$

37. *The elastic force of air at a given temperature varies inversely as the volume which it occupies.*

This law, which is generally called Boyle's Law, after its discoverer, is verified by the following experiment.

Let a tube, bent so that its two branches AB and BC are parallel, be partially filled with mercury, and placed so that each branch is vertical; the mercury will then stand at the same level, DE, in each of them; let the extremity C of the branch BC be now closed; the pressure at every point of the air thus shut in CE is uniform (Art. 35), and of course equal to the atmospheric pressure; now let more mercury be poured into the open end A; this will cause the surface of the mercury to rise both in BA and BC, but unequally: the point F, to which it ascends in BC, being much lower than G, which it attains in AB. It is always found upon ascertaining the volumes CF and CE, which may be done by weighing the mercury which they would separately contain, and upon measuring the vertical length FG, that if h be the height of the barometer observed at the time of making the experiment,

$$\frac{h+FG}{h} = \frac{\text{vol. } CE}{\text{vol. } CF}.$$

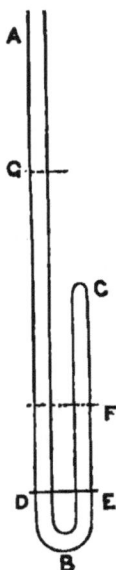

But if σ be the specific gravity of the mercury, Π the pressure of air in CE before the additional mercury was poured in, since, as above remarked, this must equal the pressure of the atmosphere,

$$\Pi = \sigma h.$$

Also if Π' be the pressure of the air when compressed into the space CF, since this must be the same as that of the mercury at the level F in the tube, and therefore equal to the atmospheric pressure at G, together with the weight of the column FG,

$$\Pi' = \sigma h + \sigma FG = \sigma (h + FG),$$

$$\therefore \frac{\Pi'}{\Pi} = \frac{h + FG}{h}$$

$$= \frac{\text{vol. } CE}{\text{vol. } CF}.$$

Therefore when *compressed* the elastic force of air varies inversely as the space which it occupies.

Next let the experiment be so far altered that instead of the step of pouring more mercury into A, a portion of the mercury already in the tube be removed; F and G may, as before, represent the surfaces of the mercury in BC and BA respectively, but in this case F will be higher than G, and both of them lower than DE.

Upon measuring, as before, it will now be found that

$$\frac{h-FG}{h} = \frac{\text{vol. } CE}{\text{vol. } CF}.$$

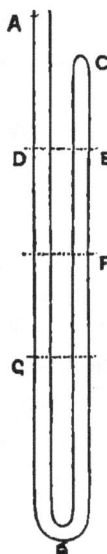

Also Π and Π' having the same meaning as before,

$$\Pi = \sigma h\ ;$$

and because Π' is the pressure at F, and therefore less than the atmospheric pressure at G by the weight of the column of mercury FG,

$$\Pi' = \sigma\,(h - GF),$$

$$\text{hence } \frac{\Pi'}{\Pi} = \frac{\text{vol. } CE}{\text{vol. } CF}.$$

And therefore the law enunciated holds equally whether the air is compressed or expanded.

It can in a similar way be verified for all other elastic fluids.

38. As the density of the same quantity of an elastic fluid varies *inversely* as the space which it occupies, we may put the above law into a more convenient form, and say, that the pressure at any point within a portion of uniform and elastic fluid varies as the density of that portion; or, since it

is quite unimportant how large this portion may be, we arrive at the general conclusion, that if p be the pressure at any point of an elastic fluid, and ρ the density of the fluid at that point, then

$$p = k\rho,$$

where k is constant for the same fluid at the same temperature; its value may be ascertained by experiment.

The formula which connects p and ρ when the temperature of the fluid varies will be the subject of investigation in Section V.

Also, for a description of the Thermometer and its use in measuring changes of temperature, see the same Section.

38*.   To find k for air.

Let p be the pressure and ρ the density of the air at a place where the barometer stands at h inches, ρ' the density of mercury.

Then, (Art. 38) $p = k\rho,$

by the barometer $p = g\rho'h,$

$$\therefore k = g . \frac{\rho'}{\rho} . h.$$

The ratio $\frac{\rho'}{\rho}$ may be found by any method for determining the specific gravities of gases and solids. The numerical value of g, for given units of time and space, is known by experiments with the pendulum: the value of h in terms of the unit of length assumed in the value of g is observed. Hence the numerical value of k for air is ascertained. A similar method will serve for any gas, the pressure of which can be determined by a barometer.

It must be remembered that the value of k thus found belongs to p, referred to the square on the assumed unit of length as unit of area, and expressed in terms of the weight of

$\left(\dfrac{1}{g}\right)^{\text{th}}$ of a cube of the substance by reference to which ρ and ρ' are estimated, and of which each edge is the unit of length.

39. By the aid of the result of Art. 38, we may discover the law of variation of the density of the atmosphere in reference to the elevation above the earth's surface, supposing the temperature to be constant, and the force of gravity to be the same as at the earth's surface.

Suppose the atmosphere up to any proposed height z feet from the earth, to be divided into a great number of horizontal layers of equal thickness τ; by taking this number (which may be represented by n) large enough, τ may be made so small that the density throughout each layer may be considered approximately uniform, and equal to that at its lowest surface; let ρ_s represent generally the density, p_s the pressure at the lowest surface of the sth layer, reckoning upwards from the earth's surface, k the known constant proportion for air at the given temperature between the pressure and the density: then (Art. 38)

$$p_s = k\rho_s,$$
$$p_{s-1} = k\rho_{s-1}.$$

Now consider the equilibrium of a small vertical prism of the sth layer of horizontal section α, and height τ: the vertical forces on it will be the pressures on its end and its weight.

$$\therefore\ p_{s-1}\alpha = p_s\alpha + g\rho_{s-1}\tau\alpha,\ \text{nearly,}$$

or
$$p_{s-1} - p_s = g\rho_{s-1}\tau,\ \text{nearly;}$$

$\therefore$ by substitution,

$$k(\rho_{s-1} - \rho_s) = g\tau\rho_{s-1},$$
$$\text{or } \frac{\rho_s}{\rho_{s-1}} = 1 - \frac{g\tau}{k}$$

a ratio which is constant and less than unity, and therefore
the densities of the successive layers, proceeding upwards
from the earth's surface, form the terms of a decreasing
geometric series.

If z' be the height above the earth's surface of the top of
the n'^{th} layer, a convenient form may be found for comparing
the densities at the two heights z and z', which will indicate
a means of finding by the aid of the barometer the difference
between them.

Writing down the ratios of all the successive densities
between z and z' we have

$$\frac{\rho_n}{\rho_{n-1}} = 1 - \frac{g\tau}{k} \ \ldots\ldots\ldots\ldots\ldots (1),$$

$$\frac{\rho_{n-1}}{\rho_{n-2}} = 1 - \frac{g\tau}{k} \ \ldots\ldots\ldots\ldots (2),$$

$$\&c. = \&c.$$

$$\frac{\rho_{n'+1}}{\rho_{n'}} = 1 - \frac{g\tau}{k} \ \ldots\ldots\ldots (n - n');$$

$$\therefore \frac{\rho_n}{\rho_{n'}} = \left(1 - \frac{g\tau}{k}\right)^{(n-n')},$$

but $z = n\tau,$

$$z' = n'\tau;$$

$$\therefore n - n' = \frac{z - z'}{\tau},$$

and, by substitution,

$$\frac{\rho_n}{\rho_{n'}} = \left(1 - \frac{g\tau}{k}\right)^{\frac{z-z'}{\tau}}.$$

The smaller τ is made, the more nearly does our reasoning
approach the real case, and therefore the more nearly will this
result be true.

But putting for convenience $z - z'$, the difference between the two heights, equal to x, and expanding,

$$\left(1 - \frac{g\tau}{k}\right)^{\frac{x}{\tau}} = 1 - \frac{x}{\tau}\frac{g\tau}{k} + \frac{\frac{x}{\tau}\left(\frac{x}{\tau} - 1\right)}{1.2}\left(\frac{g\tau}{k}\right)^2 - \&c.$$

$$= 1 - \frac{gx}{k} + \frac{1 - \frac{\tau}{x}}{1.2}\frac{g^2x^2}{k^2} - \frac{\left(1 - \frac{\tau}{x}\right)\left(1 - 2\frac{\tau}{x}\right)}{1.2.3}\frac{g^3x^3}{k^3} + \&c.$$

which, as τ is made smaller and smaller and approaches zero, approaches to the value

$$1 - \frac{gx}{k} + \frac{1}{1.2}\frac{g^2x^2}{k^2} - \frac{1}{1.2.3}\frac{g^3x^3}{k^3} + \&c.,$$

that is, the more nearly our supposed case approaches the real one, the more nearly true is the equation

$$\frac{\rho_n}{\rho_{n'}} = 1 - \frac{gx}{k} + \frac{1}{1.2}\frac{g^2x^2}{k^2} - \&c.$$

$$= \varepsilon^{-\frac{gx}{k}}$$

$$= \varepsilon^{-\frac{g(z-z')}{k}}.$$

We may therefore take this to be the true value of the ratio between the densities of the air at two heights, z and z' feet above the earth's surface, upon the supposition of gravity and temperature being both constant.

40. If h and h' be the observed heights of the barometer at places whose elevations above the earth's surface are z and z', these must be proportional to the pressures, and therefore to the densities of the atmosphere at those places; hence

$$\frac{\rho_n}{\rho_{n'}} = \frac{h}{h'} = \varepsilon^{-\frac{g(z-z')}{k}};$$

$$\therefore \ z - z' = \frac{k}{g} \log_\epsilon \left(\frac{h'}{h}\right),$$

a formula from which the difference between the heights above the earth of the two places z and z' may be readily found.

It need hardly be observed that the lengths z, z', τ, h, h' are all necessarily estimated in terms of the same unit.

In practically finding the height of a place by barometrical observation, the variation in temperature and in gravity cannot be neglected; the hygrometrical state, too, of the atmosphere must be considered; for these reasons the preceding formula cannot be confidently made use of when great accuracy is required, but the method by which it was obtained sufficiently well illustrates the principles followed in the general case.

Ex. At the base of a mountain the barometer stands at 30 inches; on the summit at 25 inches; the ratio of the density of mercury to the density of air at the base is 10,000: find the height of the mountain, given that

$$\log_{10} 6 = \cdot 7781, \quad \log_{10} 5 = \cdot 6989, \quad \log_\epsilon 10 = 2 \cdot 3025.$$

By the formula,

$$\text{the height required} = \frac{k}{g} \log_\epsilon \frac{30}{25}$$

$$= \frac{k}{g} \log_\epsilon 10 . \log_{10} \frac{6}{5}$$

$$= \frac{k}{g} 2 \cdot 3025 \times (\cdot 7781 - \cdot 6989)$$

$$= \frac{k}{g} \times 2 \cdot 3025 \times \cdot 0792$$

$$= \frac{k}{g} \times \cdot 1823.$$

Now the pressure of air at the base $= k\rho = g\rho'h$;

$$\therefore \frac{k}{g} = \frac{\rho'}{\rho} h = 10000 \times 30,$$

if an inch be the unit of length;

$$\therefore \text{height} = \frac{10000 \times 30 \times \cdot 1823}{12} \text{ feet}$$

$$= 4557\tfrac{1}{2} \text{ feet.}$$

41. The instruments whose description fills up the remainder of this Section are some of those whose action depends upon hydrostatical principles.

The Air-pump

is employed to exhaust the air from a closed vessel called a receiver. There are many modifications of this instrument, but the principles upon which they all depend, and the parts of them essential to their working, are illustrated by the annexed figure.

A is the receiver, gene-rally a large glass vessel, having its edges ground very smooth; it is placed upon a polished platform, through which it communicates by the tube *HC* with the cylinder *B*; *DE* is a piston closely fitting this cylinder and worked by the rod *G*; in the piston *DE* and at the extremity of the tube *HC* are the valves *F* and *C*, both opening outwards.

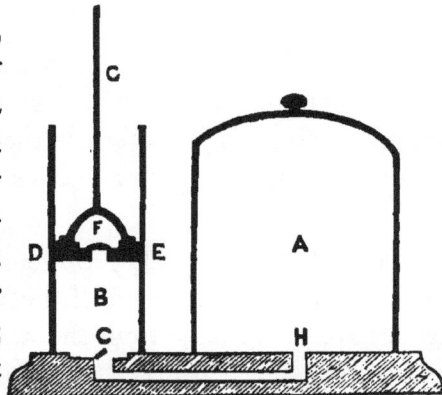

Suppose the receiver A to be full of air at atmospheric pressure, and the piston DE to be at the bottom of the cylinder B: next suppose DE to be drawn up by the rod G; as it rises the valve F must remain shut, for it will be pressed in by the external atmosphere, therefore no air can enter B from without, and C being thus free from any downward pressure, will be opened by the pressure from the air within the cylinder A; hence this air will flow into B, and when DE gets to the top of the cylinder, the quantity of air which at first filled A alone will fill A and B together; and therefore the quantity which is now in A, is to the quantity which was there at first, as is $A : A + B$, A and B representing the volumes of A and B. Now suppose DE to descend; C immediately shuts, and the air in B being compressed by the descent of DE, overcomes the pressure of the external air upon F, and therefore opens the valve and escapes, the air in A remaining undisturbed; hence when the piston has returned to its first position at the bottom of the cylinder, or, as it is usually termed, has completed its stroke,

quantity of air in A at the end of stroke

$$= \frac{A}{A + B} \times \text{(quantity there at beginning)}.$$

By a repetition of the strokes, the quantity of air in A may be diminished in the same proportion every time, and by proceeding long enough, although we cannot reduce it to absolute zero, we may make it as small as we like.

42. If Q, Q_1, ... Q_n represent the mass of the air in A originally, and at the end of the first, second, and n^{th} strokes respectively, we have from the above reasoning

$$Q_1 = \frac{A}{A + B} Q \quad \dots\dots\dots\dots (1).$$

$$Q_2 = \frac{A}{A+B} Q_1 \cdots\cdots\cdots\cdots (2),$$

&c. = &c.

$$Q_n = \frac{A}{A+B} Q_{n-1} \cdots\cdots\cdots\cdots (n);$$

∴ by multiplication,

$$Q_n = \left(\frac{A}{A+B}\right)^n Q.$$

Also, if $\rho, \rho_1 \ldots \rho_n$ represent the densities of the air in A at these times respectively, since the density varies as the mass in the same volume,

$$\rho_n = \left(\frac{A}{A+B}\right)^n \rho.$$

43. It will be remarked that, for the effective working of this instrument, the valve F must open as DE descends, and shut when it rises, while just the reverse must be the case with C. The first will clearly be insured if DE be made to fit very closely to the bottom of the cylinder, for by that means, however small the quantity of air in B, as DE descends, it will always at the end of the stroke be so compressed that its elastic force upwards shall exceed the pressure of the external atmosphere upon F together with F's own weight, which are the two forces tending to keep F shut; and it is manifest that as DE rises, the first of these two forces acting downwards on F is greater than that of the rarefied air in A and B, and hence F will never open at this stage. As regards C, when DE first moves towards the top of the cylinder, there will be no air to press it downwards from B, and the sole force acting upon it from that direction will be its own weight; if therefore the elasticity of the air left in A can overcome this, the valve must open and admit a portion of the air into B,

which will be very approximately of the same density as that left in *A*, the only check to the equalization being the weight of *C*: hence clearly there will be no difficulty about *C*'s shutting as *D* descends. It thus appears that the weight of *C* is the only cause which limits the amount of exhaustion capable of being produced in *A*. In considering the action of the valves friction ought not perhaps to be neglected, but in the kind of valve most generally used, which is merely a square or triangular piece of oiled silk fastened by its corners over a wire grating, it seems to be reduced to an extremely small amount.

44. The force, required in an instrument of the above construction, to draw *up* the piston, is the difference between the pressure downwards of the external atmosphere upon *DE* and that of the rarefied air in *B* upon the same surface upwards: at every stroke this difference increases and soon becomes very considerable; of course the same force acts in aid of the downward stroke, when it is not wanted: *Hawksbee's* air-pump is distinguished by a contrivance which · neutralizes these forces: the annexed figure represents a portion of it; two cylinders *B* and *B'* of exactly the same dimensions and provided with valves and piston-rods, as before described, communicate by pipes leading from the lower valves *C* with the same receiver *H*. Both piston-rods are worked by the same crank and toothed wheel, and consequently as one ascends the other descends; the atmospheric pressure therefore which retards the ascending one is exactly balanced by that which accelerates the descent of the other.

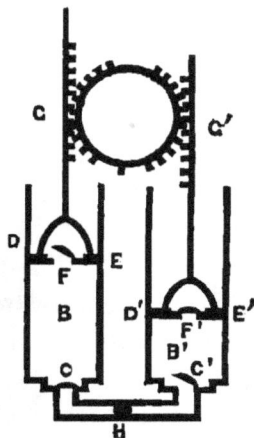

45. In *Smeaton's* air-pump one cylinder only is used, but this is provided with a valve K at its upper extremity opening outwards: the effect of this is to take off the external atmospheric pressure during a part of the stroke.

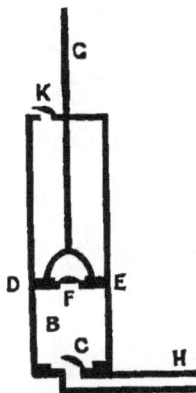

46. *The Condenser,*

as its name denotes, is employed to force air into a vessel or receiver up to any required density. The annexed figure represents it.

A is the receiver, generally a very strong hollow copper sphere; B is a hollow cylinder within which a piston DE works, carrying a valve F which opens downwards: B and A communicate by means of a pipe, at the orifice of which is a valve C also opening downwards; the piston DE is worked up and down by means of the rod G.

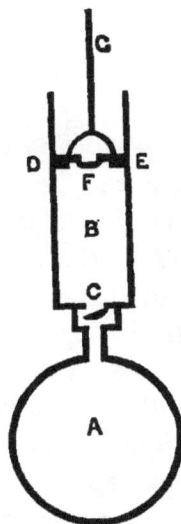

Suppose A and B to be filled with air at atmospheric pressure, then as DE descends, F shuts and C opens, and thus the air which was initially in B is forced into A: when DE begins to rise the pressure of the air in A shuts C, so that during the whole ascent the quantity of air in A remains unaltered, but meanwhile F opens, and at the end of the stroke B is again as at first full of air at atmospheric pressure; hence it is clear that by a repetition of this process, at every complete stroke a quantity of air equal to that contained by B at atmospheric pressure will be forced into A. Hence if A and B represent the volumes of A and B respectively, and if Q be the quantity of air estimated by its mass, which is contained in A at atmospheric pressure, then, since $\dfrac{B}{A}\,Q$ will be what B contains at atmospheric pressure, after n strokes A will contain the quantity $Q + n\dfrac{B}{A}\,Q$: or, representing this by Q_n,

$$Q_n = Q\left(1 + n\frac{B}{A}\right).$$

If ρ_n and ρ represent the corresponding densities,

$$\frac{\rho_n}{\rho} = 1 + n\frac{B}{A}.$$

The Common Pump.

47. The construction of this pump is explained by the annexed figure: AB is a cylinder in which a tight fitting piston EF is worked by means of a rod H; EF carries a valve G which opens upwards, and at the bottom of the cylinder is a valve D also opening upwards, and covering the orifice of a pipe D which communicates with a reservoir of water at I: BC is the highest range of the piston, and K is an open spout.

For the explanation of the working of this pump, suppose EF to be at A the bottom of the cylinder; then when it begins to rise, there being no longer any downward pressure upon D, the air in DI will open it by its elasticity and will flow into the space between A and EF; it will thus become

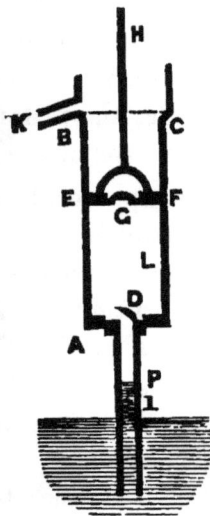

rarefied, and therefore its pressure upon the water within the pipe DI will be less than that of the atmosphere upon the external surface of the water; the water will consequently be forced a little way up the pipe, until the pressure due to the rarefied air and this column of water is equal to that of the atmosphere. Since the air in the pipe becomes more and more rarefied as EF ascends, until it has attained its greatest height BC, the column of water in the pipe will all this time be continually increasing. When EF descends, D will shut, and, as in the air-pump, the air between EF and A will be gradually condensed until it opens the valve G, and entirely escapes by it, while the condition of the air and water in the pipe DI will remain unchanged. When EF has returned to A the stroke is completed, and it is easy to see that a repetition of it will cause the water to rise gradually higher and higher in DI, until at length it will enter the cylinder BA: the next rise of the piston will, of course, remove all the air remaining above the surface of the water in BA together with some of the water itself, and ever after, provided AD be not higher above the surface of the water I than the height of the column of water required to balance the atmosphere, the water in the pipe will follow the rise of the piston up to the same level in BA, and hence at every successive stroke the piston which returns through this without disturbing it will lift out at the spout K the quantity of water contained above A.

The cause which makes the water rise in the pump is, as appears by the above explanation, the excess of the atmospheric pressure upon the surface of the external water, above the pressure of the rarefied air upon the internal water, and therefore the extreme limit to which this internal water can rise is the height of the column in the water barometer, or about 32 feet: but it is clearly necessary, for even a partial working of the machine, that the free surface of the water within the pipe should come above A; hence it is essential that the lower part of the cylinder of a pump be at a distance less than 32 feet from the surface of the reservoir, from which it is required to raise the water.

If the piston in its range does not descend to D, let L be the lowest point of its descent. Then, unless the air occupying BD when the piston is at its highest, gives, when condensed into LD, a pressure greater than that of the atmosphere the valve G will not open. Therefore not only must L be, as we have already seen, within 32 feet vertically above the surface of the water in the reservoir, but also it must not be so near that limit, that any stroke of the piston before the water reaches it shall reduce the air in BD below the just mentioned density.

48. If α be the area of the piston EF, the force employed by means of the rod H to raise it at any stroke, is, omitting the consideration of friction, equal to the difference between the pressure upon α of the external air downwards and the pressure upon α of the internal rarefied air upwards; now if σ be the specific gravity of water, h the height of the water barometer, the pressure of the air is σh: also if P be the height of the water in the pipe at the time of the stroke, the pressure of the rarefied air at the piston being the same as that at P, must be the same as that externally at I diminished by that due to the column PI; it therefore equals

$$\sigma h - \sigma PI = \sigma (h - PI),$$

therefore pressure downwards upon piston $= \sigma h \alpha$,

.... upwards .... $= \sigma (h - PI)\alpha$,

$\therefore$ the tension of the rod which is the difference between these $= \sigma PI \alpha$

$=$ weight of column of water whose base is EF and height PI.

49. *The Lifting Pump*

is the same as the common pump, except that the rod *H* plays through a water-tight socket, and the spout *K* is replaced by a pipe of any required length, provided at its junction with the cylinder with a valve *L*, which opens outwards. It is evident that as the piston descends this valve will shut and prevent the water, raised into the pipe *K* by the previous ascent of the piston, from returning, and hence every stroke will *lift* more water into *K* until it be raised to any required height.

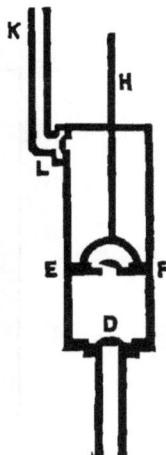

50. *The Forcing Pump*

is another modification of the pump, by which water may be raised to any height. In this case the piston *EF* contains no valve, but is quite solid: at the bottom of the cylinder *BA* enters a pipe *MN* of any length whatever, provided with a valve *M* which opens outwards. At the descent of the piston *D* shuts and the water between *EF* and *A* is *forced* up the pipe *MN*; upon the rise of the piston the return of the water from *MN* is prevented by the valve *M*.

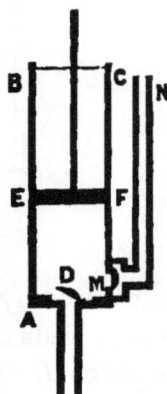

Bramah's Press.

51. The annexed figure represents this machine:

A is a very large solid cylinder or piston, working freely through a water-tight collar *EF* into a hollow cylinder *EFGH*;

A supports a large platform *BC*, which is carried up or down by the ascent or descent of *A*. At *H* is a pipe whose orifice

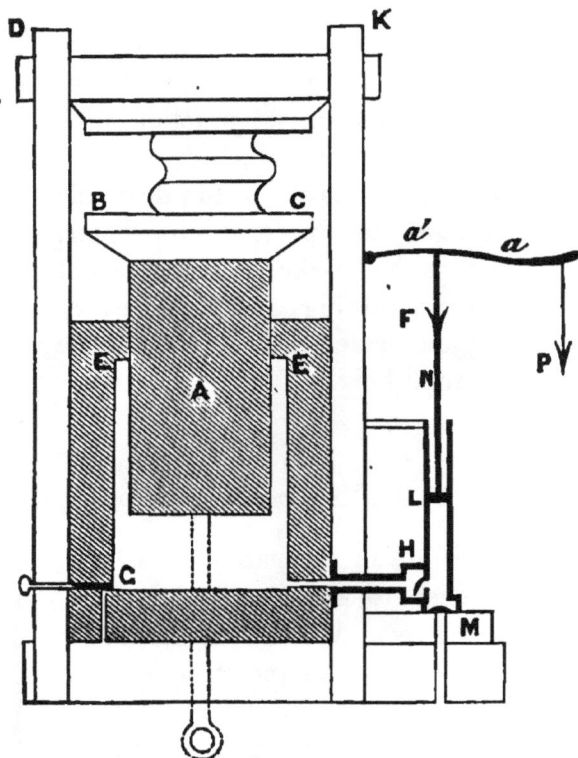

is covered by a valve opening into the large cylinder *GH* and which leads into a smaller cylinder *LM*: in this cylinder works a piston *L* by means of a rod *N*, and at the bottom is a pipe leading to a reservoir of water and covered by a valve *M* which opens into the cylinder *LM*.

Suppose both cylinders to be filled with water and the valve *M* to be closed; if then a force be applied to the piston *L*, it will be transmitted through the fluid to all surfaces in contact with it, and therefore to the lower surface of *A*; by this means A will be pushed upwards, and any substance placed upon the platform may be pressed against a fixed framework DK; when L has arrived at the bottom of the cylinder it can be drawn back, the water in GH will then be prevented from returning by the valve H, and water will also be pumped through M into the cylinder LM; L may now again be forced down, and therefore A be raised higher, until the substance between BC and DK has been sufficiently compressed. This machine may also be employed for the purpose of producing tension, in rods and chains, &c., by rigidly attaching a piston-rod to the lower part of A, which should pass through a water-tight collar in the bottom of the cylinder GF, and carry a ring at its outer extremity, to which the rod or chain to be strained may be connected.

The pressure exerted by A may at any instant be removed by unscrewing a cock at G, by which means the water is allowed to escape.

Let W represent the force with which A is pressed upwards, during a stroke of L downwards, made under the action of a force F; we may consider these forces as just balancing one another, and the pressure at all points of the water to be uniform, as the effect of gravity may be neglected in comparison with the pressure transmitted from F and W; let this be represented by p; if then r be the radius of the lower end of A, r' the radius of the surface of L, we have

$$W = \text{pressure on end of } A = p\pi r^2,$$
$$F = \text{pressure} \ldots\ldots\ldots\ldots L = p\pi r'^2;$$
$$\therefore \frac{F}{W} = \frac{r'^2}{r^2}.$$

If, moreover, as is usually the case, F is produced by the aid of a lever, whose arms are represented by a and a', and the power by P,

$$\frac{P}{F} = \frac{a'}{a} \; ;$$

$$\therefore \; \frac{P}{W} = \frac{a'}{a} \frac{r'^2}{r^2}.$$

This ratio may be rendered excessively small by reducing a' and r' as compared with a and r, and the only limit to the enormous force which this machine may be made to exert, is put by the strength of the materials of which it is framed. It will appear, on considering the result of Art. 69, that the larger cylinder is required to be the stronger.

The Siphon.

52. If a bent tube, as *ABC*, in the annexed figure, be filled with water, and then, both its extremities being closed to prevent the escape of the water, if it be inverted and one of its ends *A* immersed in a vessel of water whose surface is exposed to atmospheric pressure, while the other *C* remains outside the vessel at a level lower than the surface *D* of the water in the vessel, and if, when in this position, the ends *A* and *C* be opened, the water will be observed to flow continuously from the vessel, along the tube and out at the extremity *C*, until the surface *D* has been lowered to the level of *C* or of *A*, if *C* be lower than *A*. A bent tube so employed is called a *siphon*.

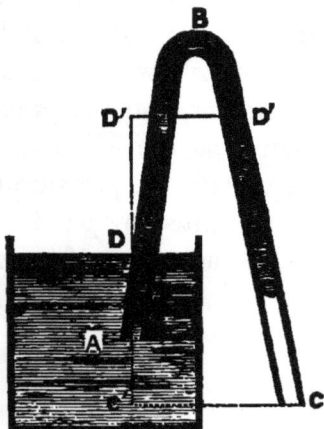

The explanation of this phenomenon is as follows: If in the side of a cylinder GH containing water, an orifice F be made below the level of the free surface GE, the water will of course flow out; for the particles of fluid inside at F are pressed outwards with a fluid pressure equal to the pressure of the atmosphere at the level GE together with that due to the height EF of *water*, while they are pressed inwards by the pressure of the air at F equal to the pressure of the atmosphere at the level E together with that due to the height EF of *air;* it is therefore on the whole pressed *outwards* with a force proportional to the depth EF, the proportion being the difference between the specific gravities of water and air. The same would be true for a vessel of any shape, because, whenever a fluid is continuous throughout a vessel, and is acted upon by gravity alone, the pressures at the same level below the surface are the same in every portion of the fluid. In the first figure of this Article, the water in the tube and in the vessel forms one continuous mass, hence the pressure at all points in the same level, wherever taken, must be the same, and therefore the fluid pressure in the tube at C must be the same as the fluid pressure in the vessel at C', C and C' being in the same horizontal line: if then C be lower than the free surface D, the water will flow out at C: by the removal of each particle of water, all resistance to the motion of the next behind it, under the pressure to which it is subjected, disappears, and thus a continuous stream will be produced towards C.

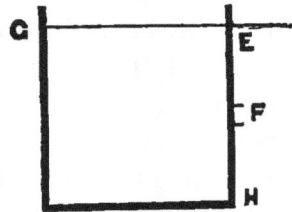

If we further consider the pressure at different points in the tube ABC, for all those which are in the leg BC *below* the level of D, it will, from what is said above, be greater than that of the outside atmosphere, and therefore if an aperture were made at any such point, the water would flow out as it does at C; for all those, however, which are above this level in either leg, as for instance in any level D', it must be less than that at D by the amount of the pressure due to the weight of the intervening column $D'D$ of water, while the pressure of the exterior atmosphere in the same level differs from that at D by the weight of the same height of air only: hence the pressure of the fluid in the tube at the level D' is less than the corresponding pressure of the external air, and if therefore an aperture were made at any such point D', the air would flow in and drive the water out along each leg of the tube.

The highest point B of the tube must not be more than 32 feet above the free surface D of the water in the vessel, otherwise the water in AB would not be supported.

53.　Perhaps the action of the siphon may be advantageously illustrated by the consideration of the following case.

$ABB'A'$ is a bent uniform tube with both branches parallel and vertical; it is occupied by a fluid (say water or air), and the whole is surrounded by another fluid (say air or water); the two fluids are kept separate, if necessary, in the tubes by very thin laminæ C and C', whose areas will be that of the section of the tube $= a$ suppose.

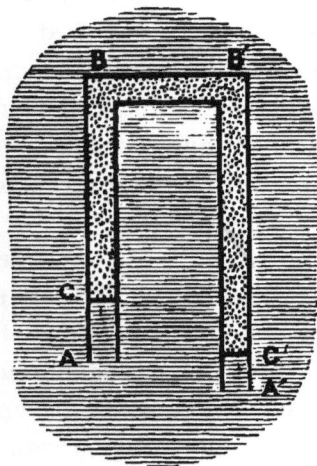

Let Π represent the pressure at the level BB' in the external fluid, σ its specific gravity, σ' that of the fluid in the tube: Π' the pressure of the tube BB' vertically downwards upon the included fluid: then the pressure at the depth C in the external fluid is $\Pi + \sigma BC$, while at the same depth in the internal fluid it is $\Pi' + \sigma'BC$, hence the resultant pressure upwards upon the surface C, which is the difference between these two over each unit of area of α

$$= \{\Pi - \Pi' + BC(\sigma - \sigma')\}\, \alpha = P \text{ for shortness.}$$

Similarly the pressure upwards upon the surface C'

$$= \{\Pi - \Pi' + B'C'(\sigma - \sigma')\}\, \alpha = P'.$$

Supposing Π to be so large that each of these expressions is positive when $\sigma - \sigma'$ is negative, we see that these pressures will always balance each other when $BC = B'C'$, but that when one leg as $B'C'$ is longer than the other BC, then P is greater or less than P', according as $\sigma - \sigma'$ is negative or positive: thus if the interior fluid be water and the exterior air, as in the case of the siphon, σ is less than σ', and therefore P is greater than P', and the water will be pushed round from the shorter to the longer leg: if however the interior fluid be air and the outer water, σ being greater than σ', P' is greater than P, and therefore the air will be pushed round from the longer to the shorter leg.

If Π were too small to make P and P' positive when $\sigma - \sigma'$ is negative, the resultant pressures upon C and C' would be downwards, and the fluid in the tubes would flow out until its height in each was just enough to make P and P' zero: this supposed case is analogous to that of the siphon when B is more than 32 feet from the surface D.

The Diving Bell.

54. Suppose a heavy hollow cylinder as $ABCD$, open at CD and closed at the top AB, to be lowered by a rope EF out of air into water; when the mouth CD is at the surface of the water the vessel would be full of air at atmospheric pressure, as it descends this air becomes compressed into a smaller space and the water rises into the vessel; but if seats be affixed within it at a sufficient height, persons

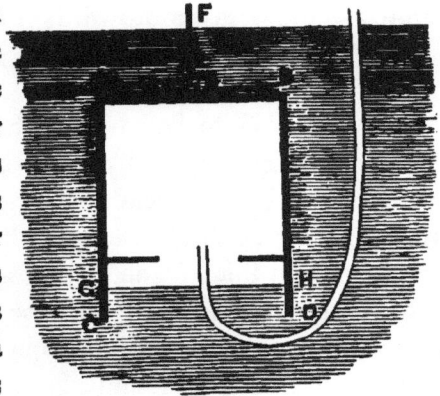

seated upon them might by this means safely descend to a considerable depth; such a contrivance is called a Diving Bell; the object of its being open at the bottom is to afford means of access to external subjects of investigation.

[NOTE. The weight of the bell must be greater than the weight of the volume of water displaced by the inclosed air when AB is level with the upper surface of the water.]

The height to which the water rises in the bell, when its lower extremity D has descended to a depth d below the surface of the water, may be found without difficulty.

Let GH be the level required of the water in the bell, and call DH, z: the atmospheric pressure at the surface of the water may be represented by Π, the specific gravity of the water by σ, and the altitude BD of the bell by a: then the pressure of water at depth H, which is $d-z$, must be

$$\Pi + \sigma\,(d-z).$$

Since this is counteracted by the pressure of the compressed air, which is to that of the atmosphere :: vol. AD : vol. AH, we must have

$$\frac{\Pi + \sigma(d - z)}{\Pi} = \frac{\text{vol. } AD}{\text{vol. } AH} = \frac{a}{a - z},$$

a quadratic equation from which z may be determined.

The tension of the string FE is equal to the weight of the bell and the inclosed air, minus that of the water displaced; or if T be this tension, W the weight of the bell, w that of the inclosed air, and A be the area of the horizontal section of the cylinder,

$$T = W + w - \sigma A . BH.$$

In practice a flexible tube is passed down to the bell under its lower edge, and air is forced through it from a condenser, so that the surface of the water GH is kept at any desired level and not allowed to rise to an inconvenient height. Other tubes provided with valves are also employed, by the aid of which the air may be changed when it is unfit for respiration.

The Atmospheric Engine.*

55. The annexed diagram represents a section of the Atmospheric Engine invented by Newcomen in the year 1705, for working the pumps of mines.

AB is a massive wooden beam (turning about an axis C, strongly supported), having its extremities terminated by circular arcs, which are connected by chains, the one (A) to the pump-rods (loaded, if necessary), the other to the rod of

---

* For the following description of the Atmospheric and Double-Action Steam Engine I am indebted to the kindness of a friend.

a solid piston *E* working steam-tight in an accurately bored cylinder *F*. This cylinder is open at the top, but closed at

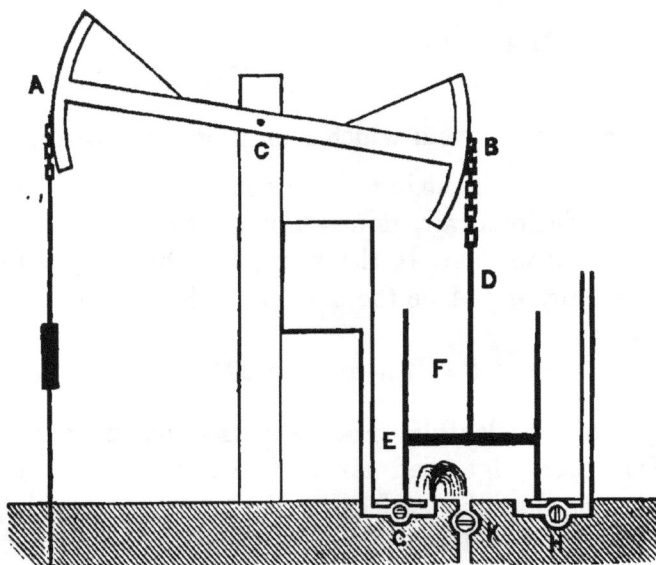

the bottom, in which are the orifices of three tubes, *H, G, K*, each furnished with a cock, and passing, respectively, into the boiler, an elevated cistern of cold water and a waste pipe. Suppose steam to be generated in the boiler, the three cocks all closed and the piston at the bottom of the cylinder, it will be kept in this position by the atmospheric pressure on its upper surface. Let the cock *H* be opened, steam from the boiler will enter the cylinder below the piston and counterbalance the atmospheric pressure on its upper surface. The weight of the pump-rods being now unsupported will depress the extremity *A* of the beam and raise the piston to the top of the cylinder: the cock *H* is now closed and *G* opened, through which a jet of cold water rushing into the cylinder condenses the steam and forms a vacuum, more or less perfect, below the piston which is now driven down by the atmospheric pressure on its upper surface, raising in its descent the pump-rods connected with *A* together with their load of water; the cock *G* is now closed, and the condensed steam and water let off by the cock *K* which is then also closed. The machine has now completed one stroke and is in the same condition as at first. Hence the operation may be repeated at pleasure.

The cocks *H*, *G*, *K*, originally turned by hand, were, by a contrivance of a youth named Potter, afterwards worked by the machine itself.

Watt's Improvements.

56. Such was the engine which came under the observation of James Watt, whose comprehensive genius perceiving its various defects, suggested amendments so complete as to bring it almost to the perfection of the beautiful engines of the present day. The following is an outline of his most material improvements.

The source of the motive power is the heat which is applied to the water in the boiler, and which calls into play the elastic force of the steam; and the real expense of working the machine is caused by the consumption of fuel required for generating this heat. It occurred to Watt that a great useless expenditure of heat was entailed by the foregoing method of condensing the steam at the end of each stroke; for while it is only wanted to cool the steam itself, the jet of cold water evidently lowered the temperature of the cylinder also, and therefore caused it every time to abstract a portion of the heat of the newly-introduced steam. To obviate this he added a separate vessel (the condenser), in which the operation of condensing the steam might be performed, so that the cylinder should remain constantly of the same temperature. A further saving was also effected by pumping the contents of the condenser back to the boiler.

He now closed the cylinder at the top, and admitting steam alternately above and below the piston, converted the atmospheric into the double-action steam-engine.

He also invented that beautiful contrivance, the parallel motion, for keeping the extremity of the piston-rod in the same vertical line, while the end of the beam with which it is connected describes an arc of a circle.

The Double-action Condensing Engine.

57. The annexed diagram represents a section of a double-action condensing Engine, in its simplest form.

A is a tube by which steam is conveyed from the boiler to the steam-box *B*, which is a closed chamber having its side adjacent to the cylinder truly flat. In this side are the apertures of three tubes *E*, *F*, *G*, of which the two former enter the cylinder at the top and bottom respectively; the third passes into the condenser *L*; *C* is the slide-valve, being a piece of metal having one side accurately flat so as to slide in steam-tight contact with the flat face of the steam-box. In this face of the slide is cut a groove in length not greater than the distance between the apertures *E*, *F*, diminished by the width of the aperture, as in fig. (2). The rod of the slide-valve passes through a steam-tight collar in the bottom of the steam-box, and is connected by a lever *OP* and a rod *PQ* with the beam of the engine, by which it receives a motion the reverse of that of the piston-rods. *D* is the cylinder

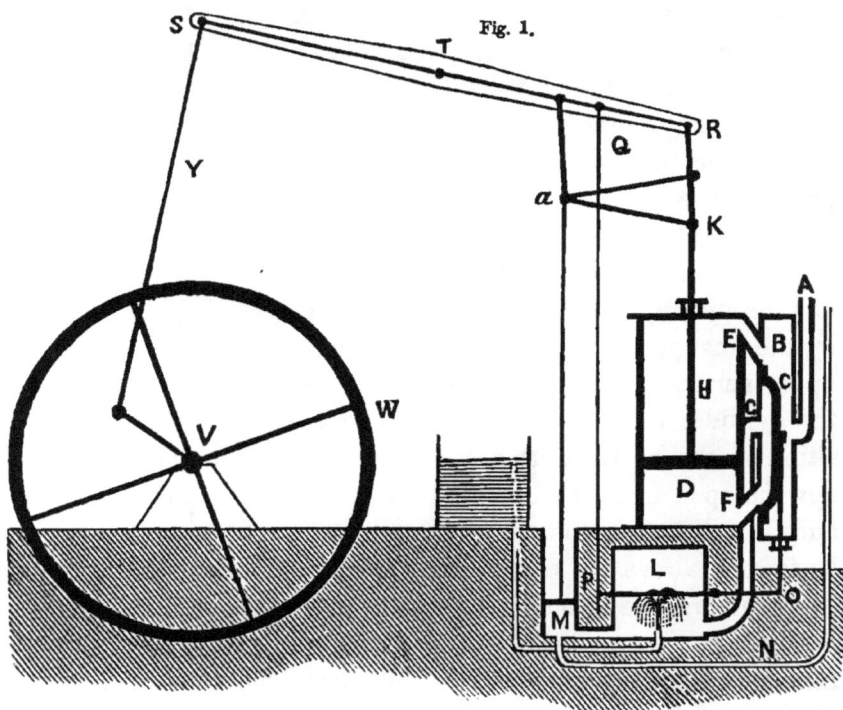

Fig. 1.

closed at both ends, truly bored and accurately fitted with a solid piston whose rod *H* works in a steam-tight collar in the top of the cylinder, and is connected with the beam by the parallel motion *KaR*. *RS* is the beam turning about an axis *T* and having its extremity *S* connected by a rod and crank with an axle on which is the fly-wheel *W*. *L* is the condenser, a closed vessel in which a jet of cold water is constantly playing. *M* is a force-pump, worked by a rod connected to the beam, which returns the condensed steam and water from the condenser by the tube *N* to the boiler.

Suppose the piston at the top of the cylin- Fig. 2.
der, the lower part filled with steam, the slide-
valve being in the position shewn in fig. (2).
A passage is now open by which steam will
pass from the boiler through the tube *A*, the
steam-box *B*, and the tube *E* into the upper
part of the cylinder, and another by which the
steam from the lower part of the cylinder will
pass by the tube *F*, the groove of the slide,
and the tube *G* to the condenser, and being
there condensed a vacuum will be formed below the piston, which will now be driven down
by the unsupported pressure of the steam on its
upper surface.   When the piston has reached the bottom of
the cylinder the slide-valve will have been shifted to the position shewn by the dotted part of fig. (2).   The passage will
now be open from the boiler to the lower part of the cylinder,
and from the upper part to the condenser; a vacuum will thus
be formed above the piston, which will then be driven up by
the pressure of the steam below it, and when the piston has
reached the top of the cylinder the slide-valve will have
reassumed its original position, the engine has now made one
stroke, and its parts are in the same positions as at first; the
operation will therefore continue.

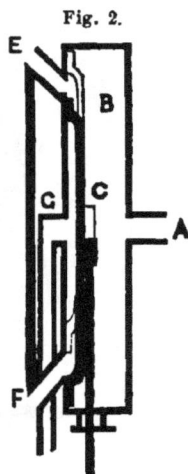

Remarks.

It will be observed that the force exerted by the engine
tending to produce a rotatory motion in the axle *V*, is greatest
when the arm of a crank is in a position at right angles to the
direction of its connecting rod *Y*, and that it diminishes to
nothing as the arm revolves to a position in which it coincides
in direction with it.   The object of the fly-wheel is by its

momentum to convert this variable into a constant force, and also to prevent the mischief which might arise from shocks or sudden variations, in the resistance to be overcome.

In practice the steam is frequently shut off from the cylinder when the piston has performed half or even less of its stroke; by this means fuel is saved, and the motion is rendered more even and smooth.

EXAMPLES TO SECTION III.

(1) A piston fits closely in a cylinder, a length a of the cylinder below the piston contains air at atmospheric pressure: compare the forces sufficient to hold the piston drawn out through a distance b with that sufficient to maintain it pushed in through the same distance.

Let Π represent the atmospheric pressure, and A the area of the piston, then at first the pressure upon the piston, both of the external and the internal air, is ΠA; by the compression through the space b the volume of the internal air is diminished in the ratio of a to $a-b$; and therefore by Boyle's law its pressure is increased in the same ratio: hence the difference between the internal and external pressures over the area of the piston is in this case $\Pi A \dfrac{a}{a-b} - \Pi A$, or $\Pi A \dfrac{b}{a-b}$: the force, required to maintain the piston so far pushed in, is of course equal to this.

Similarly, by pulling the piston out through a distance b, the volume of the internal air is increased, and therefore its pressure diminished in the ratio of $a+b:a$; hence the difference between the external and internal pressures over the area of the cylinder, which is the same as the force required to hold it in this position, is now

$$\Pi A - \Pi A \frac{a}{a+b}, \quad \text{or} \quad \Pi A \frac{b}{a+b}.$$

The ratio asked for between these two forces is manifestly

$$\frac{a+b}{a-b}.$$

(2) If a cylinder be full of air at atmospheric pressure, and a close-fitting piston be forced in through $\frac{1}{3}$ of the length of the cylinder by a weight of 10 lbs., shew that it will require an additional weight of 30 lbs. to force it through $\frac{2}{3}$ of the length.

If Π be the pressure, estimated in pounds, of the air in the cylinder at first, $i.\,e.$ be the pressure of the atmosphere, the pressure after the given compression into $\frac{2}{3}$ of the volume is, by Boyle's law, $\frac{3}{2}\Pi$; therefore if A be the area of the piston, the pressure upon it upwards is $\frac{3}{2}\Pi A$, and by question this must balance the pressure of the external atmosphere together with 10 lbs. acting downwards, hence

$$\frac{3}{2}\Pi A = \Pi A + 10,$$
$$\text{or } \Pi A = 20 \text{ lbs.};$$
$$\text{and therefore } \frac{3}{2}\Pi A = 30 \text{ lbs.}$$

This pressure will be doubled, by the same law, when the compression is extended to another $\frac{1}{3}$ of the cylinder; therefore the compressing force must be 60 lbs.; hence 30 lbs. must be added to that which is already acting. Q. E. D.

(3) A bubble of gas ascends through a fluid, whose free surface is open to the atmosphere, and whose specific gravity is s; supposing the bubble to be always a small sphere, compare its diameters, when at depths d and d', having given that

the height of the barometer is h and the specific gravity of mercury σ.

Since the bubble is always small, the pressure of the fluid upon its surface may be considered to be everywhere the same, and to be that which is due to the depth of the center of the bubble; also since this pressure is the only force limiting the volume of the bubble, it must be exactly equal to the elastic force of the air forming the bubble.

Now the pressure of the fluid at the depth d is $\sigma h + sd$,

.... $d' \ldots \sigma h + sd'$;

and the elastic force of the air in the bubble varies inversely as its volume, that is, inversely as the cube of the diameter; hence if a, a' be the diameters corresponding to the depths of d, d', we must have

$$\frac{a'^3}{a^3} = \frac{\sigma h + sd}{\sigma h + sd'},$$

which gives the required ratio between the diameters.

(4) An imperfect barometer is compared at two different times with a true one, and it is found that the readings h_1, h_2 are less than the true readings by the quantities ϵ_1, ϵ_2 respectively. Shew that the true reading may at any other time be obtained by adding to the observed reading h the correction

$$\frac{\epsilon_1 \epsilon_2 (h_2 - h_1)}{\epsilon_1 (h - h_1) + \epsilon_2 (h_2 - h)}.$$

The error in the height of the mercury must be due to the presence of a certain quantity of air in the upper part of the tube, whose elastic pressure supplies the place of the weight of the deficient length of mercury; this pressure will not be constant, but will continually alter, for, the quantity of air remaining the same, its volume diminishes as the barometer rises, and increases as it falls.

Let x_1, x_2, x be the length of tube occupied by it when the barometer is at height h_1, h_2, and h respectively, ρ_1, ρ_2, ρ its corresponding densities, and k the proportion between the pressure and the density of air, (Art. 38); then, by question, the pressure of this air, when included in the length x_1, must be equivalent to the weight of a length ϵ_1 of mercury; hence if σ be the specific gravity of mercury,

$$\left.\begin{array}{l} k\rho_1 = \sigma\epsilon_1 \\ \text{similarly} \quad k\rho_2 = \sigma\epsilon_2 \\ k\rho = \sigma\epsilon \end{array}\right\} \quad \cdots\cdots\cdots\cdots (A).$$

And since the volume of the air remains the same, the transverse section of the tube being constant,

$$\rho_1 x_1 = \rho_2 x_2 = \rho x \cdots\cdots\cdots\cdots (B).$$

Also since the length occupied by air, together with that occupied by mercury, must always make up the whole length of tube above the zero point,

$$h_1 + x_1 = h_2 + x_2 = h + x \cdots\cdots\cdots (C).$$

These seven equations will enable us to determine the seven unknown quantities ρ_1, ρ_2, ρ, x_1, x_2, x, ϵ; but as ϵ is the only one of these which we want, it will be convenient to eliminate all the others.

Multiplying the equations (A) by x_1, x_2, x respectively, and then comparing them with those of (B), we find

$$\epsilon_1 x_1 = \epsilon_2 x_2 = \epsilon x = \lambda, \text{ suppose};$$

therefore, substituting these values of x_1, x_2, x in (C), we get

$$h_1 + \frac{1}{\epsilon_1}\lambda = h_2 + \frac{1}{\epsilon_2}\lambda = h + \frac{1}{\epsilon}\lambda = \mu, \text{ suppose,}$$

or more conveniently,

$$h_1 + \frac{1}{\epsilon_1}\lambda - \mu = 0 \ldots\ldots\ldots (1),$$

$$h_2 + \frac{1}{\epsilon_2}\lambda - \mu = 0 \ldots\ldots\ldots (2),$$

$$h + \frac{1}{\epsilon}\lambda - \mu = 0 \ldots\ldots\ldots (3).$$

Multiplying (1) by $(h_2 - h)$, (2) by $(h - h_1)$, (3) by $(h_1 - h_2)$, and adding the resulting equations, we obtain

$$\frac{1}{\epsilon_1}(h_2 - h) - \frac{1}{\epsilon_2}(h - h_1) + \frac{1}{\epsilon}(h_1 - h_2) = 0,$$

and therefore $\epsilon = \dfrac{\epsilon_1 \epsilon_2 (h_2 - h_1)}{\epsilon_1 (h - h_1) + \epsilon_2 (h_2 - h)}.$

(5) If a barometer be standing at the height of 30 inches and be placed under the receiver of an air-pump in which the capacity of the barrel and receiver is the same, what will be the height of the mercury after three strokes of the pump?

The height of the barometer is directly proportional to the pressure of the air upon it and therefore to the density of that air; now since the capacity of the receiver is the same as that of the barrel, the density of the air in it is diminished one half at each stroke of the piston (Art. 42); therefore at the end of the third stroke it will be $\frac{1}{8}$th of what it was at first, *i. e.* of atmospheric density, and, consequently, the corresponding height of the barometer will be $\frac{1}{8}$th of 30 inches, or $3\frac{3}{4}$ inches.

(6) The receiver of an air-pump has 20 times the volume which the barrel has, and a piece of bladder is placed over a hole in the top of it: the bladder is able to bear a pressure of 3 lbs. the square inch, and the pressure of the atmosphere is 15 lbs., shew that the bladder will burst between the 4th and 5th strokes. Given that $\log 2 = \cdot 30103$, $\log 21 = 1 \cdot 3222193$.

The resultant pressure upon the bladder is the difference between the pressure of the air outside and that of the air inside the receiver; hence, by question, the bladder will burst when this difference becomes as great as 3 lbs. the square inch; *i. e.* since the pressure of air varies as its density, when the density of the inside air is to that of the atmosphere as 12 : 15, or as 4 : 5.

Now after the nth stroke the density of the inside air equals $\left(\dfrac{20}{21}\right)^n$ of that of the atmosphere (Art. 42), therefore the bladder will burst during the stroke, whose number being put for n is the first whole number which makes $\left(\dfrac{20}{21}\right)^n$ less than $\dfrac{4}{5}$, or which makes

$$n\{1 + \log 2 - \log 21\} \text{ less than } 3\log 2 - 1;$$

this number may by the aid of the given logarithms be shewn to be 4.

(7)　A cylinder whose length is 3 feet is fitted with an air-tight piston: when the piston is forced down to within 9 inches of the bottom what will be the pressure of the air within, supposing the pressure at first to have been 15 lbs. on the square inch?

(8)　A cylinder whose base equals a square foot and whose height is 7 inches is filled with common air whose pressure may be assumed to be 14 lbs. per. square inch and covered with a moveable lid without weight; shew that if a weight of 336 lbs. be placed on the lid it will sink 1 inch.

(9)　A bubble of air one foot below the surface of a pond had a diameter of one inch: what was its diameter when it was 8 feet below, neglecting the pressure of the atmosphere?

(10)   The barometer in ascending a mountain sinks from 29 to 23 inches: find the change in the pressure on each square inch, the specific gravity of mercury being 14.

(11)   A closed cylinder with its axis vertical is filled with two gases which are separated by a heavy piston.   Determine the position of the piston, it being given that either fluid, if it filled the whole cylinder, would support a pressure equal to $\left(\dfrac{3}{8}\right)^{\text{ths}}$ of the weight of the piston.

(12)   A tube of uniform bore, open at one end, is fixed with its axis vertical and open end upwards: shew that, unless its height be greater than that of the barometric column, no amount of mercury, so poured in that no air escapes, will produce equilibrium.

(13)   When the mercury in a barometer stands at 30 inches, shew that the pressure of the atmosphere on a square inch is about $14\frac{3}{4}$ lbs., the specific gravity of mercury being 13.6, and the weight of a cubic foot of water 1000 ozs.

(14)   A barometer has the area of the cistern four times that of the tube, and when the mercury stands at 30 inches, 2 inches of the tube remain unfilled: if a mass of air, which at the density of the atmosphere would fill one inch of the tube, were admitted into the upper portion, shew that the column would be depressed 4 inches.

(15)   Compare roughly the masses of the atmosphere and the earth: given that, the mean height of the barometer is 30 inches, the density of mercury is 13, the mean density of the earth is 5, and the radius of the earth is 4000 miles.

(16)   Given that the specific gravity of air at the earth's surface at a given place is ·00125, and of mercury is 13, when the barometer is at 30 inches; find the height of the homogeneous atmosphere.

(17)   If the volume of the receiver of an air-pump be six times that of the piston-cylinder, and if a barometer introduced into the receiver stand at 28 inches after one ascent of the piston, at what height will it stand after two more ascents of the piston?

(18)   Supposing the upper valve of Smeaton's air-pump to open when the piston is half-way up, what was the density of air in the receiver at the beginning of the ascent?

(19)   If the volume of the receiver of an air-pump be ten times that of the barrel, shew that before the eighth motion of the piston is completed, the density of the air in the receiver will be reduced one-half, having given $\log 2 = .30103$, $\log 11 = 1.0413927$.

(20)   If the capacity of the receiver of a condenser be ten times that of the barrel, after how many descents of the piston will the force of the condensed air be doubled?

(21)   If a pump be employed to raise a fluid whose specific gravity is .915, find the greatest distance which is admissible between the lower valve and the surface of the fluid.

(22)   In what position must a siphon ABC (having angle ABC a right angle, and $AB = BC$) be placed in a vessel full of water, so as to empty from it the greatest quantity possible?

(23)   Find the limit to the height of the highest point of a siphon above the surface of water at a place where the mercurial barometer stands at 25 inches, the specific gravity of mercury being 13.58.

(24)   Two equal cylinders containing equal quantities V of different fluids (of specific gravities $\sigma_1 \sigma_2$), which will not mix, are connected by an exhausted siphon of small bore,

which reaches to the bottom of each cylinder; find how much fluid will run from one vessel into the other.

(25)   Does the tension of the rope of the diving-bell increase or decrease as the bell is lowered?

(26)   Find the position of unstable equilibrium of a light diving-bell.

(27)   Find the volume of air at atmospheric density which may be pumped into the bell at a given depth.

(28)   Two diving-bells are suspended at the ends of a rope which passes over a smooth wheel; find how much the one must be heavier than the other, that they may rest in a given position.

What would happen if the rope were hollow, so as to form a tube of communication between them?

(29)   If a valve be made in the top of a diving-bell opening upwards, what will be the result?

(30)   In the hydraulic press used at the Britannia bridge, the internal diameter of the great cylinder was 20 inches, and that of the small tube $1\frac{1}{8}$ inch; find the lifting power of the press, supposing a pressure of 8 tons to be exerted by the piston in the small tube.

SECTION IV.

58. The Statical principle of Virtual Velocities asserts that if any forces as $R_1 R_2 \ldots R_n$ acting upon a system of points keep each other in equilibrium, and if r_1, $r_2 \ldots r_n$ be the virtual velocities of these forces respectively, consequent upon any very small displacement of the system, which does not alter the molecular connexion of its different points, then

$$R_1 r_1 + R_2 r_2 + \ldots + R_n r_n = 0.$$

The same principle can be proved to hold, when $R_1 R_2$ &c. instead of being applied to points of a rigid system, act normally upon the surfaces of a weightless inelastic fluid, the condition of displacement being in this case that the volume of the fluid remain constant and continuous.

Let $\alpha_1 \alpha_2 \ldots \ldots \alpha_n$ be the portions of surfaces, taken small enough to be considered plane, upon which R_1 &c. $\ldots R_n$ respectively act; then since these forces are in equilibrium, the pressure per unit of surface must be the same throughout (Art. 7), and if it be represented by p, we shall have

$$\frac{R_1}{\alpha_1} = \frac{R_2}{\alpha_2} = \&c. \ldots \ldots \frac{R_n}{\alpha_n} = p \ldots \ldots \ldots (1).$$

Suppose the displacement of α_1 &c. to be made by moving them respectively through distances r_1, $r_2 \ldots r_n$ along tubes perpendicular to them and closely fitting them; these distances will be proportional to the virtual velocity of the corresponding force; also the alteration in volume of the including vessel which the movement of each area α has thus produced in αr, and is an increase or decrease according as r is positive or negative, and therefore the total alteration will be the sum of these quantities, each r being supposed affected with its proper sign; but since by the supposition the disturbance was not such as to break the continuity of the fluid or to alter its volume, whatever space was gained in the displacement of any of the pistons, an equal space must have been lost in the displacement of some of the others, so that by this means the total alteration is nothing; therefore we must have

$$a_1 r_1 + a_2 r_2 + \ldots\ldots + a_n r_n = 0 \ldots\ldots\ldots (2),$$

and hence, by substitution from equations (1),

$$R_1 r_1 + R_2 r_2 + \ldots\ldots + R_n r_n = 0 \ldots\ldots\ldots (3).$$

If the fluid be heavy the principle will still be true, and can be proved upon the assumption that normal resistance is the only mutual action between the particles. And even in this case, if $R_1 R_2 \ldots R_n$ represent respectively the excess of the forces acting upon the several pistons $a_1 a_2 \ldots a_n$ beyond the forces which would be required to balance the effect of gravity alone, since then equations (1) (see Art. 7) and (2) will still hold, the truth of (3) will be established as before.

59. If a heavy fluid be contained in a vessel and the pressures on its sides be resolved in three directions at right angles to each other, to find the sum of the resolved parts in each direction.

Let HLK represent any such vessel, HK being the level of the fluid contained in it; let AA' be a line in the fluid drawn parallel to the direction of one of the required resolved forces, to meet the sides in A and A'.

Let a very small cylinder be constructed, having AA' for its axis, ω for its transverse section, a and a' the sections of its extremities made by the sides of the vessel, and therefore portions of those sides, these will be approximately plane. Let P, P' be the normal pressures of the fluid upon these, not generally in one plane, and let θ, θ' be the angles which they respectively make with AA'; then resolving P and P'

along AA' and perpendicular to it, if R be the resultant of the resolved parts of P and P' along AA',

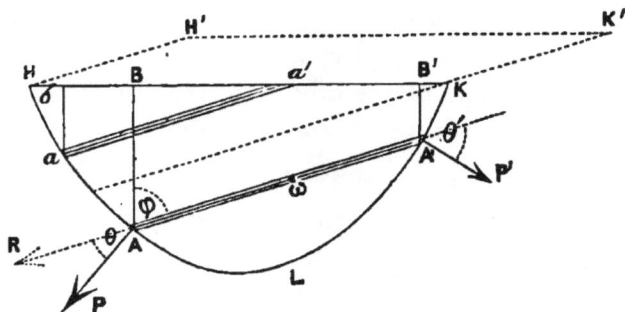

$$R = P\cos\theta - P'\cos\theta';$$

but if Π be the pressure of the atmosphere, or any other pressure which is uniform throughout the surface HK, and σ be the specific gravity of the fluid,

$$P = (\Pi + \sigma AB)\alpha, \qquad P' = (\Pi + \sigma A'B')\alpha';$$
$$\therefore R = \Pi(\alpha\cos\theta - \alpha'\cos\theta') + \sigma(AB\alpha\cos\theta - A'B'\alpha'\cos\theta').$$

Also $\alpha\cos\theta = \omega = \alpha'\cos\theta'$, therefore this expression is reduced to

$$R = \sigma\omega(AB - A'B').$$

If ϕ be the angle which AA' makes with the vertical,

$$AB - A'B' = AA'\cos\phi;$$
$$\therefore R = \sigma\omega AA'\cos\phi.$$

But $\sigma\omega AA'$ is the weight of the column of fluid AA', let it be represented by w, then

$$R = w\cos\phi \ldots\ldots\ldots\ldots (I).$$

This result is true for all the columns parallel to AA' which have both their extremities meeting the sides of the vessel; let aa' in the figure be a column, one extremity only of which, i. e. a, meets the side, the other a' is made by the surface of the fluid, and let the same letters, as above, be used with the same meanings in this case, then $P' = 0$, and $\theta' = \phi$;

$$\therefore R = P\cos\theta = (\Pi + \sigma ab)a\cos\theta$$
$$= \Pi a' \cos\theta' + \sigma ab\omega$$
$$= \Pi a' \cos\phi + \sigma\omega aa' \cos\phi$$
$$= (\Pi a' + w)\cos\phi \dots\dots\dots \text{(II)}.$$

Now the magnitude of the resultant required of the resolved forces is the sum of all the R's for which (I) holds, together with the sum of all the R's for which (II) holds, *i. e.* it is the sum of *all* the quantities ($w\cos\phi$), together with the sum of the quantities $\Pi a' \cos\phi$ as long as a' is the section of the cylinders made by HK; but the sum of all these sections is the whole area of the surface itself, let it be represented by A, and the sum of all the w's is the weight of the fluid contained in the vessel, let this be W: we thus come to the conclusion, that the *magnitude* of the resultant of the resolved parts in directions parallel to AA' of the fluid-pressures upon the sides of the vessel containing the fluid is

$$(\Pi A + W)\cos\phi \dots\dots\dots\dots \text{(III)}$$

where ϕ is the inclination of AA' to the vertical; the other resolved parts of the pressures are in a plane perpendicular to AA'.

60. Also this force being the resultant of a set of parallel forces, each of which acts along the axis of a column of the fluid as AA', and is proportional to the magnitude of that column (for even the R in (II) may be considered as proportional to the magnitude of aa' increased by a constant quantity), its *direction* must pass through the center of gravity of the mass, which is made up of these columns, *i. e.* in the case under consideration, the resultant of the resolved pressures parallel to AA' must pass through the center of gravity of a mass of fluid represented by $HLKK'H'$, the portion $H'HKK'$ being made up of the portions which must be added to the columns aa' in order to make up the pressure Π in the formula (II).

If Π be zero, it appears that the resultant of each resolved part will always pass through the center of gravity of the fluid HLK.

61. If instead of the preceding case we had considered a body HLK immersed in a fluid whose free surface was HK produced, we should have obtained the same result as (III) for the *magnitude* of the resolved pressure in the direction parallel to AA', but its direction would have been reversed.

62. It is easy to deduce from (III) that, if the pressures be resolved in directions vertical and horizontal, the horizontal portions vanish while the vertical one equals $\Pi A + W$: this, then, is the magnitude of the total resultant of all the pressures, which accords with Art. (18).

63. If the fluid contained in HLK were without weight, the pressure at every point of its surface would be Π, which might be supposed to be the result of any such cause as that mentioned above, or to be produced by the elasticity of the fluid itself: in the latter case, of course, HK could not be open.

Under these circumstances, if the fluid pressure upon the surface HLK be resolved in any direction AA', whose inclination to the plane HK which cuts off this surface, is $\frac{\pi}{2} - \phi$, its resultant is by the preceding investigations

$$= \Pi A \cos \phi,$$

where A is, as before, the area of the plane HK.

Or, this result may be stated more generally as follows, so as to include the cases where the boundary of the surface HLK is not a plane section.

If the surface HLK *be projected upon a plane perpendicular to* AA', *then the resolved part in the direction* AA' *of a normal pressure over the surface* HLK, *which is uniform and equal to*

Π at every point, is the same as would be produced if the pro-
jected area were pressed uniformly with a pressure Π.

64. DEF. When any surface is submitted to fluid pressures, the point in it, where the direction of the resultant of
these pressures would meet it, is called the *Center of Pressure;*
if the pressures are such that they have no *single* resultant,
there is no center of pressure.

In general, to determine this point it is necessary to employ the Integral Calculus, but in some simple cases of fluid
pressures on plane surfaces its position may be inferred from
a knowledge of the position of the center of gravity of some
solid.

Take the case of any plane surface immersed vertically in
a fluid, which is at rest under the action of gravity alone:
suppose this surface divided into horizontal strips, each so
narrow that the pressure on it may ultimately be considered
uniform.

Then the pressure upon each strip will be equal to the
weight of a rectangular slice of the fluid, which has the strip
for its base, and the depth of the strip below the surface for its
height (Art. 18).

Therefore the condition of the pressed surface will be the
same as if it were placed horizontally with the just mentioned
slices of fluid (considered rigid for the purpose) standing upon
the corresponding strips of surface.

These slices in the aggregate form a solidified mass of
fluid, and the resultant of their weights is the weight of this
mass acting through its center of gravity.

Therefore the pressure on the surface immersed is equal to
the weight of the solid, constructed as just mentioned, and
acts through its center of gravity in a direction normal to the
surface.

By construction this solid is always the frustrum, made by a plane, of the right prism which has the surface immersed for its base and the greatest depth of such surface for its greatest height (see also Art. 18).

Thus if the surface immersed be a triangle with its base in the surface of the fluid, the solid of pressure is a pyramid having the triangle for its base and the altitude of the triangle for the edge which passes through the vertex. The depth of the center of gravity of this pyramid below the surface of the fluid is easily ascertained to be $\frac{1}{2}$ the altitude of the triangle immersed: the required center of pressure will be therefore at the same depth and in the line joining the vertex of the triangle with the middle point of its base.

65. Let CD be a tube, capable of revolving freely about a vertical axis, and having at its lower extremity C one or more arms CA, CB, &c. horizontal and closed at their ends A, B, &c.: suppose this to be filled with water and then orifices to be opened in CA, CB, one in each, and always on the same side looking from the center C; as the water flows out at these orifices the whole instrument will revolve in the opposite

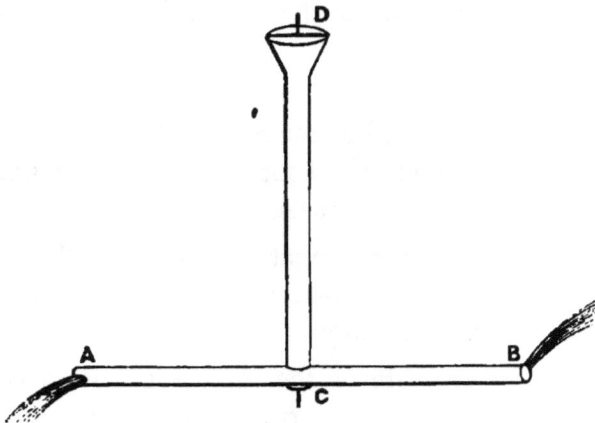

direction about the axis CD. Such an instrument is usually termed *Barker's Mill*.

The explanation of the cause of the revolution is very simple; the fluid may be divided into a number of horizontal cylinders, each perpendicular to the axis of *AB*, and the resolved part along these of the pressure at their extremities will be zero by (I) of Art. (59), excepting for those cylinders, one of whose extremities terminates in an orifice; the pressure at the other extremity of such cylinders is not counteracted, and will therefore tend to turn the arm round the axis. By the above-mentioned arrangement of the orifices in the different arms, the uncounteracted pressures in each will all tend to turn them in the same direction, and thus a rapid revolution of the instrument will be produced. By keeping *CD* full, this rotation may be maintained for any length of time, and be applied to mechanical purposes.

66. *The surface of a heavy fluid contained in a vessel, which revolves with a uniform angular velocity about a fixed vertical axis, is a paraboloid.*

Each particle in the fluid evidently describes with a uniform velocity a horizontal circle, having its center in the vertical axis; hence, by the principles of Mechanics, the resultant accelerating force upon the particle must tend to the center, and must $= \dfrac{v^2}{r}$, where *v* is its linear velocity and *r* the radius of the circle it describes: but this resultant is the effect of gravity and the fluid-pressures only.

Now let us consider a particle *M* in the surface of the fluid, having a mass *m*, and let the next figure represent a vertical section of the fluid made through the axis *BE* of the vessel *ADFC* and the particle *M*, *AMBC* being the curve in which it cuts the fluid. Since *M* is on the surface of a fluid, which from the mobility of the particles must be perfectly smooth, the pressure of this surface upon it may be assumed to be

normal; let this direction meet the
axis in P and draw MN horizontal;
then by what precedes the fluid-
pressure in the direction MP, and
the weight of M in the direction
of the vertical must have a resultant
along MN: the three sides then of
the triangle MPN being in the directions of, must be proportional to,
these three forces: we therefore obtain

$$\frac{PN}{MN} = \frac{mg}{m\dfrac{v^2}{r}}.$$

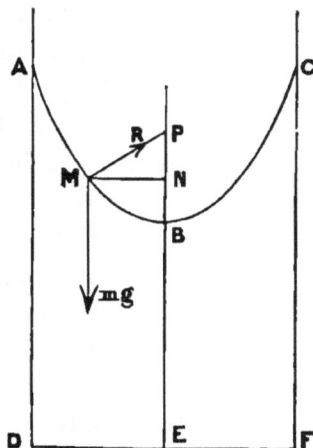

Now ω being the angular velocity of the whole vessel
about the axis NBF, expressed in terms of the angular unit
of the circular measure, v the linear velocity of M must be
ωMN; also r is the distance MN; hence the preceding relation becomes

$$\frac{PN}{MN} = \frac{g}{\omega^2 MN},$$

$$\therefore PN = \frac{g}{\omega^2},$$

that is, the subnormal PN is a constant quantity, wherever
in the curve M be taken, a result which is characteristic of
the parabola where the subnormal is always equal to $\frac{1}{2}$ the
latus rectum; the section ABC of the surface of the revolving
fluid is therefore a parabola whose latus rectum $= \frac{2}{\omega^2}g$, and,
since this section is any whatever through the axis, the whole
surface must be a paraboloid having the axis of revolution
for its axis.

It is apparent from the investigation that the form of the surface of the revolving fluid is quite independent of the shape of the including vessel.

Ex.   A cylinder full of fluid is set to revolve twice in a second about its vertical axis.   Find the quantity of fluid spilt, supposing the radius of the cylinder to be one foot.

The form of equilibrium of the free surface must be a paraboloid passing through the rim of the vessel.

Let h be the depth of the vertex of the paraboloid below the plane through the rim.   Then if r be the radius

$$r^2 = \frac{2}{\omega^2} \cdot gh,$$

but $\omega = 4\pi$, $r = 1$;

and $g = 32$, since one foot and one second are respectively the units of length and time;

$$\therefore\ 1 = \frac{2 \times 32}{(4\pi)^2} \cdot h, \text{ which gives } h \text{ in feet;}$$

$\therefore$ quantity spilt = volume of paraboloid

$$= \frac{1}{2}\pi \cdot r^2 \cdot h$$

$$= \frac{1}{2} \cdot \pi \cdot \frac{(4\pi)^2}{64}$$

$$= \frac{\pi^3}{8} \text{ cubic feet nearly.}$$

It is assumed here that the height of the cylinder is sufficient to keep the vertex of the paraboloid above the base.

67.   *If a body be immersed in a fluid, which is at rest under the action of any forces whatsoever, the resultant of the pressures upon its surface is exactly equal and opposite to the resultant of the forces which would act upon the fluid which it displaces.*

For if the body were taken away and the displaced fluid put back again and supposed to assume a solid form, equilibrium would obtain, for the mere solidification can produce no disturbance; but to maintain this equilibrium the resultant of the pressures on the surface of the solidified fluid must be exactly equal and opposite to the resultant of the forces applied to it; now these pressures are identical with those upon the immersed solid, because the surfaces are the same in both cases; hence the truth of the proposition.

68.   When fluid, contained in a vessel, exerts a pressure on the enveloping surface, this pressure has to be resisted by the cohesion of the particles forming the mass of the envelope. This exertion of the power of cohesion is called *tension*, and the intensity of its action, as in the case of every other force of resistance, will depend on the amount of pressure to be counteracted.   Every solid body admits of a tension up to a certain amount being exerted between its constituent particles. But when the forces applied require for their counteraction a tension beyond this amount, the material yields to the strain and is broken, or the vessel bursts.

To understand how the tension at a point in a solid body may be measured, let us consider for simplicity, a weight supported by a prismatic bar of iron or other material, with its axis vertical and with transverse section α. If the bar were cut through transversely in any section, the lower part would, unless otherwise supported, fall.   An action therefore takes place between the particles in contact throughout any transverse section sufficient to counteract the tendency of the weight below to pull them asunder.   This is the effect of the tenacity of the material, and is what has just been termed *tension*.   Of course this whole tension is the aggregate of the strain or tension at every point of the section, and although, for the reason given in Art. (2) for the case of *pressure*, the tension exerted by an actual point must be nothing, still, as the tension really called into play over different parts of the section may vary greatly, the conception of *tension at a point* remains. It may very well be measured, as pressure is measured in a fluid, by *the tension that would be exerted over a unit of area of a section of the mass, supposing the action between the particles in contact throughout that unit the same as at the proposed point.* If, then, w be the weight supported, t the measure of the tension at any point in any transverse section α, we have, supposing the tension uniform over α,

$$w = t \cdot \alpha.$$

If we consider the tension of the material forming a vessel, subject to any surface pressure, let k be the thickness of the side of the vessel at the point in question. Suppose a normal section of the side made at the point. Then the area of the section of length l, and uniform thickness of k, would be $k \cdot l$, and the tension over this (in a direction perpendicular to the section) would be $t \cdot kl$.

If in the vessel under consideration k be the same throughout, $t \cdot k$ is taken as the measure of the tension, and is generally denoted by T.

69. *To find the tension at any point of a cylindrical surface inclosing fluid in terms of the normal pressure of the fluid.*

The annexed figure represents a section of the cylinder having a breadth AB made by two planes perpendicular to the axis which passes through O. If AB be small enough, the tension at every point of it tending to break

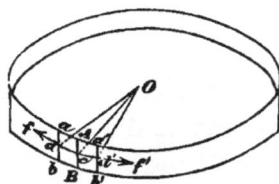

the band may be considered the same throughout; call it T; here T is, as explained in the last article, the tension that would be exerted over a normal section of the surface, of thickness equal to that of the surface, and of length equal to the linear unit, if each point of this area were solicited by a tension equal to that at the proposed point of section. Considering the thickness of the surface very small, this tension is manifestly always perpendicular to AB and in the tangent plane passing through the point at which it acts.

Draw ab, $a'b'$ parallel to AB one on each side of it, and *very close* to it, and dcd' a circular section through c the middle point of AB and parallel to aa' or bb'; then the tensions at every point of ab and $a'b'$ are very approximately equal to T, and act in their respective tangent planes. Draw the radii Od, Oc, Od'; the small area ab' is kept at rest by the normal pressures of the fluid upon it which are approximately the same at every point as that at A, and the tensions T at every point of ab and $a'b'$ acting tangentially. Hence, resolving along cO and in a plane perpendicular to it, we have

$$\text{resolved tensions along } cO$$

$$= Tab \sin cOd + Ta'b' \sin cOd', \text{ very nearly,}$$

$$= Tab \frac{dc}{cO} + Ta'b' \frac{d'c}{cO}, \text{ nearly,}$$

$$= T \frac{AB.\, aa'}{cO} \text{ nearly.}$$

Also the resolved pressures in the same direction will, if p represent the pressure at A, be

$$= p.\, aa'.\, AB \text{ very nearly:}$$

the smaller we take ab' the more nearly will these equations be true, and will be absolutely so in the limit; but in all cases these two resolved forces must be equal to one another;

$$\therefore\ p.\, aa'.\, AB = T \frac{AB.\, aa'}{cO},$$

or $T = p.cO = pr,$

if we represent the radius of the cylinder by r.

69*. This result may be obtained perhaps more simply as follows, provided that the pressure of the contained fluid throughout any plane perpendicular to the axis be supposed constant.

Suppose a section of the cylinder of breadth AB to be made by two planes perpendicular to the axis of the cylinder, and suppose this section divided into two semi-cylindrical bands by a plane through the axis. Let AB be small and k be the thickness of the material at the section. Then the plane through the axis will cut the cylindrical section in what will be, always if k be constant, *ultimately as* AB *is indefinitely diminished* if k vary, two rectangles of breadth k and length AB.

Now either of the above semi-cylindrical portions is in equilibrium under the action of the tensions between their two surfaces of junction, perpendicular to their planes, and the fluid-pressure on the concave surface. The resultant pressure parallel to the tensions will be equal to that on a rectangle of breadth AB and length equal the diameter of the cylinder (Art. 63).

Let p be the measure of the fluid-pressure, t the measure of the tension, k the thickness at the proposed section : then we have

$$2t.k.AB = p.2r.AB,$$
$$\text{or } t.k = p.r.$$

Ex. To find the law of thickness of a vessel of cylindrical bore, in order that when filled with fluid of uniform density, with its axis vertical, the tendency to burst, as measured by the tension at every point, may be the same throughout.

Let t be the tension, k the thickness at the depth z, r the radius of the bore; σ the specific gravity of the fluid.

Then, by the proposition,

$$t \,.\, k = pr,$$

but $p = \sigma z$;

$$\therefore \ k = \frac{\sigma r}{t} \,.\, z.$$

But by the question t must be constant, and σr is so,

$$\therefore \ k \propto z;$$

or the thickness must vary as the depth.

(69 A). *By a similar investigation to that of* (Art. 69) *we can find the tension at any point* A *of a spherical surface containing fluid, the fluid-pressure at that point being* p.

Take $ab\,b'a'$, a very small rectangular portion of the surface having A as its middle point, and let O be the center of the sphere, the tension at A may be resolved into two, at right angles to each other, both in the tangent plane at A; let that parallel to ab be represented by T, and that parallel to aa' by T': these may be assumed to be equal on account of the symmetry of the surface about A: then the very small surface ab' may be considered to be kept in equilibrium by the normal pressures at every point which are approximately equal to p, the tangential tensions at every point of aa' and bb' approximately equal to T, and those at every point of ab and $a'b'$ approximately equal to T'.

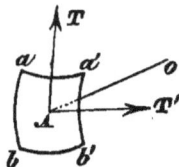

Now the resolved parts along AO of the tensions T at the points in aa' and bb'

$$= T\frac{aa'.a'b'}{AO}$$ approximately, by the last proposition,

and similarly resolved parts along AO of the tensions of ab and $a'b'$

$$= \frac{Ta'b'.aa'}{AO}$$ approximately ;

$\therefore$ the sum of these $= 2T\dfrac{aa'.a'b'}{AO}$ approximately.

Again, the resolved part of the normal pressure in the same direction

$$= paa'.a'b' \text{ very nearly.}$$

But these two resolved parts must always counteract each other; therefore, since the above expressions for them are true in the limit,

$$p = \frac{2T}{AO};$$

or if, as before, r denote the radius AO,

$$T = \tfrac{1}{2}pr.$$

(69*A). If p be the same throughout the vessel, T may be found shortly by a method similar to that of Art. (69*).

Suppose the sphere divided into two hemispherical shells by a plane through the center. Either of these is in equilibrium under the action of the tension throughout the annular band of junction, and the fluid-pressures on its concave surface. If k be the thickness of the shell, r the radius, t the measure of the tension, supposed uniform throughout, and p the pressure, we have, as in (Art. 69*),

$$t.2\pi r.k = p\pi.r^2,$$

or $t.k = \dfrac{pr}{2}.$

It must be remarked, that if the surface be pressed outside as well as within, by a fluid, the p which enters the above equations is the difference between these two normal pressures at the point A.

It appears from these formulæ, that when p is constant, T and r are proportional to each other, or that the tension which a given normal pressure calls into action at any point of a cylindrical or spherical surface containing it is proportional to the radius.

70. In the description of the Barometer given in Article 36, no explanation was offered of any method of measuring the length of the column of mercury; a fixed scale of graduations would manifestly not answer the purpose directly, as both extremities of the column necessarily shift at once: a descent of the mercury in the tube must cause the surface of the mercury in the basin to rise, and the sum of these two decrements of length makes up the total decrement; and similarly in the case of an elongation of the column. Now a fixed scale could only mark the absolute alteration at one end, say the upper C, of the column, but by a simple arrangement this may be made to indicate the total alteration. For, suppose C to sink d inches, as measured by a fixed scale; then, if a be the cross section of the tube, a volume da of mercury must thus descend into the basin, and the corresponding rise of the surface of the mercury in the basin must be due to this increase of volume; or supposing this rise to be d' inches, and the cross section of the basin to be A, we must have

$$da = d'A ;$$

hence the total diminution of the column, being, as before said, $d + d'$,

$$= d\left(1 + \frac{a}{A}\right).$$

But it is clear that if the fixed scale be divided into equal parts, each of which is to an inch as $1 : \left(1 + \dfrac{\alpha}{A}\right)$, the length d inches will contain $d\left(1 + \dfrac{\alpha}{A}\right)$ of these, and hence the number of *such graduations* observed upon the length d inches, through which C has sunk, will be the real number of *inches* by which the column has been shortened.

It only needs that the real length of the column should be practically ascertained by measuring, when C is opposite any known graduation of this scale, and be there registered, in order that all future lengths be ascertained by inspection.

71.   If the section of the tube be very small compared with that of the basin or lower vessel, $\dfrac{\alpha}{A}$ will be a small fraction, and $d + d'$ will differ inappreciably from d: this is the case with most barometers in common use, so that the graduations on their scales are made without respect to the considerations of the previous article.

72.   In some barometers the bottom of the basin containing the mercury (FG in fig. of Art. 36) is adjustable by means of a screw, and thus at the time when an observation is required to be taken, the whole mass of mercury can be raised or lowered until the surface DE is brought to a fixed level with regard to the instrument; an ivory pin, projecting from the side DF and pointing downwards, is generally used to mark this level, and the mercury can be easily brought into it by turning the screw until the image of the ivory point in the surface is made to coincide with the point itself: the fixed scale then measures upwards from this point.

73. The *Wheel* or *Siphon Barometer* differs slightly from those previously described: instead of the end *A* (fig. Art. 36) being plunged into a vessel of mercury it is bent round; the second branch so formed is similar to *BA* in the fig. of Art. (37), and has its extremity open to the atmosphere; by this arrangement the column of mercury in the second branch, together with the pressure of air on its surface, balances the column of mercury in the first; and therefore the difference between the length of these columns is the same as the *h* of Art. (36). If the tube be uniform throughout, the variation of either column is exactly half the total variation of *h*; it is therefore only necessary to observe the variation of the open one. This is done by attaching one end of a light string to a small body which is allowed to float on the surface of the mercury, and the other, after the string has been passed over a small pulley, to a weight less than that which would be sufficient to balance the float: then as the float rises and falls with the mercury, the pulley is turned round by the friction of the string, and an index needle fixed to it is made to traverse a sort of clock-face: if the circle of the face be large and carefully graduated, any very small motion of the float will be indicated and measured by the extremity of the needle.

EXAMPLES TO SECTION IV.

(1) A cylinder, the radius of whose base is 1 foot and whose weight is 100 lbs., is filled with water, a cubic foot of which weighs 1000 ozs.: if the cylinder be inverted on a smooth horizontal table, find the greatest number of revolutions per second which the water may make about the axis of the cylinder consistent with no escape of water.

Let $ABCD$ represent the cylinder inverted upon the smooth plane, and oc- cupied by fluid which is revolving about the dotted axis of the cylinder, say n times per second; its angular velocity is therefore $n.2\pi$, but we may for convenience call it ω.

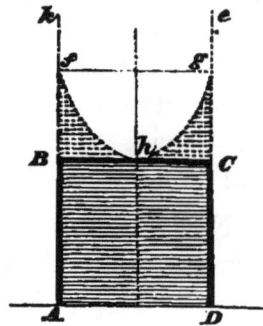

If the water were not revolving at all there would be no pressure upon BC; and again, if BC were taken away, during the revolution the water would rise at the sides until the surface took a parabolic form; BC must therefore supply some force sufficient to keep the water down, and it must be itself pressed upwards by the same: when this upward pressure becomes by reason of the magnitude of the angular velocity greater than the weight of the cylinder, i.e. 100 lbs., $ABCD$ will be lifted up and the water will escape below its edges. We want then to find this pressure corresponding to the angular velocity ω.

Suppose the sides of the cylinder to be produced upwards as represented by the dotted lines Bk, Ce in the figure, and suppose BC to be a rigid plate capable of being made to slide in and out: now let BC be drawn out while the water is revolving, and a sufficient quantity of water be added to make the curve free-surface fhg take the position given in the figure where the vertex h of the parabola is exactly where the middle point of BC was: if when this is the case BC be again pushed in, its presence can produce no disturbance in the revolving water either above it or below it; but by this means the lower part is quite shut off from the upper, which may therefore be removed and neglected; and then the lower is only under the circumstances which were proposed by the problem. We thus see that the state of the revolving fluid $ABCD$ (and therefore its pressure at every point, because it is so at one point h) is the same whether we suppose the portion of fluid $fBhCg$ to be superincumbent, or whether we suppose BC to be stretched rigidly across: hence the pressure which BC exerts downwards, and which we wanted to find, is exactly equal to the weight of the volume of fluid $fBhCg$. Call this volume V.

Now the volume of the paraboloid fhg is $\frac{1}{2}$ of that of the cylinder $fBCg$, therefore V which is the remaining half of the cylinder $=\frac{1}{2}\pi Bh^2 . Bf$. Also since the latus rectum of the parabola $fh = \frac{2g}{\omega^2}$ (Art 66), $Bf = \frac{\omega^2}{2g} Bh^2$;

$$\therefore V = \frac{\pi\omega^2}{4}\frac{Bh^4}{g}$$

$$= \frac{\pi\omega^2}{4 \times 32.2} \text{ cubic feet, because } Bh \text{ is 1 foot;}$$

$$\therefore \text{ weight of } V = \frac{\pi\omega^2}{4 \times 32.2}\frac{1000}{16}\text{lbs.}$$

This, as before explained, is the force tending to lift the cylinder, and therefore the greatest value that ω can have, without escape of water taking place, is when this just equals the weight of the cylinder, or when

$$\frac{\pi\omega^2}{4 \times 32.2} \times \frac{1000}{16} = 100,$$

$$\text{or} \quad \frac{4n^2\pi^3}{4 \times 32.2} \times \frac{10}{16} = 1,$$

$$n^2\pi^3 = 16 \times 3.22 ;$$

$$\therefore \log n = \frac{1}{2}(\log 16 + \log 3.22 - 3\log\pi)$$

$$= \frac{1}{2}(1.20412 + .50786 - 1.49145)$$

$$= \frac{1}{2} \times .22053$$

$$= .110265$$

$$= \log 1.289 ;$$

$$\therefore n = 1.289.$$

Hence the water can only make 1.289 of a revolution per second, or about $1\frac{1}{3}$, without escaping.

When would a cover, turning upon a hinge and unsymmetrically loaded, be raised?

(2) An Indian-rubber ball containing air has a radius a when the temperature of the air is $0°$ (centigrade). Supposing the tension of the Indian-rubber $= \mu \times$ (radius)^2, shew that the radius r of the ball when the temperature is $t°$ will be given by the equation

$$\frac{r^3}{a^3}\frac{\pi + 2\mu r}{\pi + 2\mu a} = 1 + et,$$

e being the ratio (Sect. V.) between the increase of volume and of temperature for air at a constant pressure, and π being the pressure of the atmosphere.

Let p and p' be the pressures of the included air, ρ and ρ' its densities when the temperature is $t°$ and $0°$ respectively, then we have (Art. 82),

$$\frac{p}{p'} = \frac{\rho}{\rho'}(1 + et).$$

Also, since the densities are inversely as the volumes,

$$\frac{\rho}{\rho'} = \frac{a^3}{r^3};$$

$$\therefore \frac{r^3}{a^3}\frac{p}{p'} = 1 + et;$$

but (Art. 69*A), $\frac{1}{2}(p - \pi)r = $ tension $= \mu r^2$ by question;

$$\therefore p = \pi + 2\mu r;$$

similarly, $p' = \pi + 2\mu a;$

∴ by substitution

$$\frac{r^3\pi + 2\mu r}{a^3\pi + 2\mu a} = 1 + et.$$

(3)   A spherical vessel contains a quantity of water whose volume is to the volume of the vessel as $n : 1$; shew that no water can escape through a small hole at the lowest point, if the vessel and the water in it have an angular velocity about the vertical diameter not less than

$$\left\{\frac{g}{r(1 - n^{\frac{1}{3}})}\right\}^{\frac{1}{2}},$$

r being the radius of the vessel, and g the accelerating force of gravity.

Let $ADBE$ be the spherical vessel revolving with an angular velocity ω about its vertical diameter ACB: the surface of the inclosed water will be a paraboloid (Art. 66), whose latus rectum is $\frac{2g}{\omega^2}$; let the upper part of this surface meet the sides of the shell in D and E: by the question no water must run out at a small hole at A, hence there must be no water lying upon the hole, or, in other words, the angular velocity is least possible when the lower part of the surface of the water just passes through A.

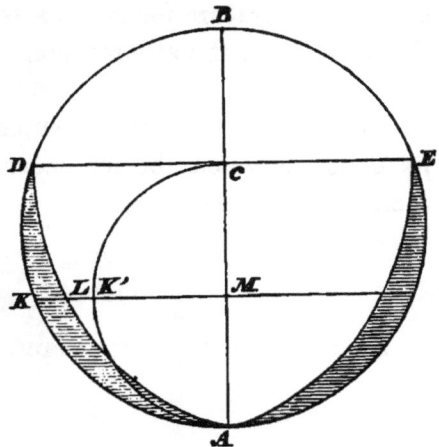

Now because E is a point in the parabola DAE, whose latus rectum is $\frac{2g}{\omega^2}$,

$$\therefore CE^2 = \frac{2g}{\omega^2} AC.$$

Also because E is a point in the circle $ADBE$, whose diameter is AB,

$$\therefore CE^2 = BC.AC;$$

$$\therefore BC = \frac{2g}{\omega^2}.$$

Again, the volume of the contained water is evidently the difference between the volume of the portion DAE of the sphere, and the volume DAE of the paraboloid; but this difference is equal to the volume of a sphere whose diameter is AC^*;

$$\therefore \frac{\text{vol. of water}}{\text{vol. of vessel containing it}} = \frac{AC^3}{AB^3} = n \text{ by question;}$$

$$\text{or} \left(\frac{AB - BC}{AB}\right)^3 = n,$$

$$1 - \frac{BC}{AB} = n^{\frac{1}{3}}.$$

---

* This is easily apparent if the slice which would be cut off from the whole sphere by any two horizontal planes very close to each other be considered, say that cut off by KLM in the figure, and a plane parallel to it at a small distance d, for the portion which belongs to the paraboloid is $\pi LM^2.d$, while the whole slice is $\pi KM^2.d$: the difference between these is $\pi(KM^2 - LM^2)d$

$$= \pi(AM.BM - AM.BC)d$$
$$= \pi AM(BM - BC)d$$
$$= \pi AM.MC.d;$$

but if $AK'C$ be a section of the sphere described upon AC as diameter $AM.MC = MK'^2$, therefore the difference between the portion of the sphere $BDKAE$, and the portion of the paraboloid DAE cut off by these two parallel planes, is equal to the slice of the sphere $CK'A$ cut off by the same planes; and this is true wherever the planes be taken, and hence the above result may be deduced.

∴ substituting from above,

$$1 - \frac{g}{\omega^2 r} = n,$$

$$\frac{g}{\omega^2 r} = 1 - n^{\frac{1}{2}}$$

$$\omega = \sqrt{\frac{g}{r\,(1 - n^{\frac{1}{2}})}}.$$

This is the value of the angular velocity which makes the surface of the revolving fluid pass through A; all less velocities would make the surface pass above A, and all greater would make the surface, produced, pass below A; hence the truth of the proposition.

(4) A bent tube of very small uniform bore throughout, consists of two straight legs, of which one is horizontal and closed at the end, and the other is vertical and open. If the horizontal leg be filled with mercury, and the tube be made to revolve about a vertical axis passing through the closed end, so that the mercury rises in the vertical leg to a distance d from the bend, shew that the angular velocity is

$$= \sqrt{\frac{2g\,(h + d)}{l^2 - d^2}},$$

l being the length of the horizontal leg, and h the height of the mercury in the barometer at the time of the experiment.

Shew also that the mercury will not rise at all in the vertical leg unless the angular velocity be greater than $\dfrac{\sqrt{2gh}}{l}$.

Let ABC represent the tube revolving about a vertical
axis through the closed end A, P the
height to which the mercury rises in
the open vertical tube BC, and Q the
distance to which it recedes from the
axis in the horizontal one, then, since
the tube is uniform, $AQ = PB = d$;
also by question $AB = l$. Suppose the
pressure of the atmospheric air upon P
to be replaced by the weight of additional mercury poured into the open
tube BC, and let the upper surface of
this mercury be at P', then $PP' = h$.

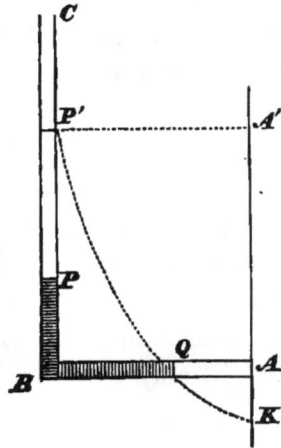

Now it is evident that since the
pressure is nothing both at P' and Q, these two points must
be in the free surface of the paraboloid, which would have
been formed by the revolution about the vertical axis through
A of a sufficient quantity of mercury contained in an open
vessel, with an angular velocity exactly equal to that of the
tube in the given case; call this velocity ω, and let the
dotted line $P'QK$ be the supposed surface: draw $P'A'$
perpendicular to the axis. Then, since the latus rectum of
the parabola $P'QK$ is (Art. 66) equal to $\frac{2g}{\omega^2}$, we have

$$AQ^2 = \frac{2g}{\omega^2} AK,$$

$$P'A'^2 = \frac{2g}{\omega^2} A'K;$$

$$\therefore P'A'^2 - AQ^2 = \frac{2g}{\omega^2} BP';$$

$$\text{or } \omega = \sqrt{\frac{2g(h+d)}{l^2 - d^2}}.$$

The least value of ω for which the mercury will rise in the tube BC is clearly that just greater than the value which makes the parabola of the surface pass through P' and A, when P will of course be at B; but under these circumstances

$$l^2 = \frac{2g}{\omega'^2}h, \text{ or } \omega' = \frac{\sqrt{2gh}}{l}. \quad \text{Q. E. D.}$$

(5)   Two hemispheres of equal radius are placed in close contact, so that their common surface is horizontal; the upper one is fixed firmly and communicates with an air-pump: find the least number of strokes of the piston in order that a given weight may be suspended from the lower hemisphere.

The internal air must be so rarefied that the difference between its normal pressure upon the lower hemisphere and that of the external air upon the same shall be great enough for its vertical resolved part (Art. 63) to be equal to the given weight, together with the weight of the lower hemisphere.

Ex.   Given the pressure of the atmosphere is 15 lbs. per square inch, the volume of the sphere is 20 times that of the pump, the area of a great circle of the sphere is 2 square feet, and the weight to be suspended, together with that of the lower hemisphere, is $19\frac{3}{5}$ cwt.

Also $\log 2 = \cdot 3010300$, $\log 2 \cdot 1 = \cdot 3222193$.

(6)   A vertical prismatic vessel, closed at the base and filled with fluid, is formed of rectangular staves held together by a single string passing round them, as a hoop.   Find the position of the string.   (Art. 64.)

(7)   A cylinder, into which water has been poured, revolves uniformly about its axis, which is vertical, bubbles of air rise from the base; shew that these will all converge towards the axis in their ascent.   (Art. 67.)

SECTION V.

MIXTURE OF GASES.—VAPOUR.

74. Boyle's law obtains for a mixture of gases as well as for a simple one; indeed, air, upon which the first experimental proof of it was practised, is made up of two gases, oxygen and nitrogen, in the proportion of 1 to 4 by weight: the experimental facts upon which the more general form of this law is based are the two following:—

(1) Whenever two gases of different densities are allowed to come into contact with each other, they very quickly intermix and form a compound which is of uniform density throughout, and in which any equal volumes, wherever taken, always contain the same proportion of the two component gases. The rapidity with which the homogeneity is attained increases with the difference between the densities of the two gases. It may here be remarked, that no such intermixture ever results from the combination of inelastic fluids unless they be of equal densities; and if it be produced among them artificially, however complete it may appear, as in milk and tea, the fluids will in time separate themselves, and lie superimposed upon one another in the order of stable equilibrium. The case of fluids, whether elastic or not, which act chemically upon one another is not here considered.

(2) If a gas compounded of given quantities of two different gases be put into a closed vessel, the pressure at every point of it, or its *elastic force* (Art. 34), is the sum of the two pressures which would be respectively due by Boyle's law to the given quantities of the two gases, if inclosed separately

P. H.   8 in the same vessel and at the same temperature as the compound: thus, let U be the capacity of the vessel, Π the pressure of the compound occupying it, and let the quantities of the two gases, forming the compound, be such that the pressure of the first when occupying a volume V is P, and of the second in a volume V' is P', then the pressure of the first in the volume U would be $P\dfrac{V}{U}$, and of the second $P'\dfrac{V'}{U}$, and therefore our result is $\Pi = P\dfrac{V}{U} + P'\dfrac{V'}{U}$, which takes the simple form

$$\Pi U = PV + P'V' \dots\dots\dots\dots (1).$$

Dr. Dalton has very concisely stated this fact by saying that " One gas acts as a vacuum with respect to another."

From this formula it can be immediately shewn that Boyle's law holds for the mixture; for supposing the same quantities of the two gases had been put into a space U' instead of U, then the pressure Π' would have been given by the equation

$$\Pi' U' = PV + P'V'.$$

Now it is manifest that this latter compound is the same as the former, and therefore Π and Π' are its pressures corresponding to the volumes U and U'; but $\Pi U = \Pi' U'$ or $\dfrac{\Pi}{\Pi'} = \dfrac{U'}{U}$, the same result as that which Boyle's law gives. (Art. 37.)

75. A gas and a liquid, when in contact with each other, and of such a character as not to act chemically upon each other, generally intermix in a partial manner: if there be several gases present a portion of each penetrates the liquid and pervades its whole extent uniformly; the amount of this portion is quite independent of the number of gases which may be so penetrating, but is always such that if it occupied

alone the volume which the liquid does, its density would bear a certain proportion to that of the same gas, which is left outside and also supposed to occupy its space alone, this proportion depending upon the liquid and the gas together.

This experimental fact may be stated more generally as follows: if the volume of a closed vessel be $V + U$, and the part U be occupied by a given liquid, and if into the remaining space V any number n of gases be introduced in any quantities whatsoever, a portion of each of them will penetrate the liquid, and the ratio which the quantity of each gas remaining in V bears to the quantity of the same which pervades the liquid in U will be independent of n: thus if $V\rho$ be the one quantity, $U\rho'$ the other for a particular gas, it is always found that $\rho' = \dfrac{1}{\mu}\rho$, whether there be only one, or whether there be fifty gases submitted to the liquid at once; as mentioned above, μ will differ with different gases, but is constant as long as the gas and the liquid which are referred to remain the same.

To take an instance: this proportion is found to be $\frac{1}{15}$ for oxygen and water, but only $\frac{1}{30}$ for nitrogen and water; hence assuming common air to consist of 1 part of oxygen and 4 of nitrogen, their densities would be in the same proportion if they occupied space alone, and therefore, by our rule, the amount of oxygen absorbed by a given portion of water in contact with air would be, when compared with the amount of nitrogen absorbed by the same portion of water, as is $\frac{1}{15} : \frac{1}{30}$ or as $1 : 2$.

76. In all our previous investigations the effect of heat in modifying the action of fluids has been left out of consideration, or rather has been supposed to be invariable: this supposition holds true whenever the circumstances of the case presume the temperature to remain constant, but if changes

take place in it, there will generally be corresponding changes in the hydrostatical properties of the fluids.

It would seem from experiment, that a free mass, whether solid or fluid, never experiences a change in its temperature without undergoing some alteration in its dimensions; and conversely. We may say generally, that additional heat imparted to a body causes it to expand, while the withdrawal of heat is followed by its contraction; the converse of this is equally general, that the forcible compression of a body makes it give out some of the heat which was necessary for its more expanded state, and the extension of it in the same way obliges it to absorb heat. Whenever, too, the temperature of any body after suffering any succession of changes returns to any particular state, it is always observed that the dimensions of the body return also to a constant corresponding magnitude.

77. By accurate observation of effects of this kind it appears also that interchanges of heat always take place between portions of matter, whether in contact or at a distance from each other, until a species of equilibrium has been attained; when this is the case the different portions are said to have the same temperature. These facts lead us to the means of estimating and defining different stages of temperature, and of measuring the amount of its increase or diminution. Any instrument constructed for this purpose is termed a *thermometer:* of these there are many kinds, but it will be sufficient, for the purpose of illustrating the above-mentioned principle, to describe the mercurial thermometer.

The Thermometer.

78. The construction of this instrument is based upon the fact, that for low temperatures the expansion of mercury under the action of heat is such that the increase of its volume is always proportional to the increase of the heat.

It is merely a glass tube, of uniform bore, developed into a bulb at one extremity; this bulb, together with a portion of the tube, is occupied by very pure mercury, and a vacuum is preserved in the remainder of the tube by the extremity being hermetically sealed; the bore of the tube is uniform and extremely small, and therefore a slight expansion of the whole mass of mercury makes a great difference in the length of the tube occupied. By the aid of graduations along the tube, the increase or decrease in the volume of the mercury consequent upon an alteration of temperature may be readily observed, and therefore the difference between temperatures compared. If the difference between two known temperatures, or any part of it, be taken as the unit of measurement of temperature, this instrument affords us the means of expressing any other difference of temperature in terms of it. The two temperatures taken for this purpose are those of melting ice, and of the steam of water boiling under an atmospheric pressure of 29·8 inches of the mercurial barometer. In some kinds of thermometers, as the centigrade, the $\frac{1}{100}$ part of the difference between these temperatures is chosen for the unit; in others, as in Fahrenheit's, the $\frac{1}{180}$ part is taken. To graduate the thermometer accordingly, the instrument must first be plunged in melting ice, and the level of the mercury in the tube marked; it must then be submitted to the steam of water boiling under a given atmospheric pressure, and the level of the mercury again marked; the volume of the tube between these two marks must be divided in the centigrade into 100, and in Fahrenheit's into 180 equal parts; each of these parts is termed a degree. These two kinds of thermometers also differ in respect to the division on the tube at which the numbering of the degrees commences; in the centigrade the reckoning starts from the freezing point, and therefore 100° indicates the temperature of boiling water, while in Fahrenheit's the initial point on the tube is 32° below the freezing point, thus making 212° the boiling water point.

These distinctions must always be carefully regarded in the consideration of the temperature as given by either thermometer; for instance, 40° centigrade indicates a temperature which is greater than freezing temperature by 40 degrees of that thermometer, (*i. e.* by $\frac{40}{100}$ of difference between freezing and boiling); but 40° Fahrenheit is only eight of its degrees (*i. e.* $\frac{1}{180}$ of difference between freezing and boiling) above freezing point: and generally, if $F°$ and $C°$ be the corresponding numbers of degrees upon Fahrenheit's and the centigrade thermometers respectively, which indicate the same given temperature, they must belong to graduations which divide the distance between the boiling and freezing point, on each thermometer, in the same proportion. Now $F°$ of Fahrenheit denotes a graduation $(F-32)$ degrees beyond freezing point, and therefore one which cuts off $\dfrac{F-32}{180}$ of the distance between boiling and freezing points, while the graduation on the centigrade whose number is C, is at a distance from its freezing point equal to $\dfrac{C}{100}$ of the length between boiling and freezing; therefore

$$\frac{F-32}{180} = \frac{C}{100}, \text{ or } 5(F-32) = 9\,C;$$

a formula connecting the numbers of those graduations on the two thermometers which correspond to the same temperature.

In a similar manner might be investigated a formula for any other two thermometers whose mode of graduation was known.

The advantage of Fahrenheit's thermometer over the centigrade and others is, that the degrees are small, and therefore fractional parts of them are the less frequently requisite in observations, and that the commencement of the scale is placed so low that it is seldom necessary to speak of negative degrees.

79. The filling and graduating of a thermometer is an affair requiring great skill and precaution. The object to be attained in filling the instrument, is to introduce a quantity of pure mercury, which shall occupy, at ordinary temperatures, the bulb and part of the tube, leaving a vacuum in the remainder.

The method generally adopted is somewhat as follows. The instrument is held vertically with its open end upwards. A small funnel-shaped vessel containing mercury is placed over this open end. The flame of a spirit lamp is applied to the bulb, which, increasing the temperature of the included air, increases its pressure, and some of it forces its way in bubbles through the mercury. If the lamp be now removed, the temperature of the air in the instrument falls, its pressure diminishes, and some of the mercury is forced in by atmospheric pressure to occupy the space of the air expelled. By continually repeating this process all the air may be dislodged and the instrument filled with mercury. If now the funnel be removed, and the instrument heated, the volume of the mercury will increase so that some will flow out at the open end. The heat is raised to the highest temperature which the thermometer can be required to indicate, and then the hitherto open end is hermetically closed. As the mercury cools, it sinks in the tube, leaving a vacuum above it.

In graduating the instrument it must be remembered that the temperature at which ice melts seems to be absolutely constant under all circumstances, but that at which water boils is not so, unless the pressure and hygrometrical state of the atmosphere is also the same.

It ought, perhaps, to be here remarked, that the expansion of the mercury measured by the thermometer is not absolutely that of the mercury itself, inasmuch as the tube too expands; it is therefore the difference between these two expansions which the graduations give us: but this fact introduces no difficulty, because the expansion of the glass, like that of the mercury, is proportional to the increase of heat, and therefore the difference between the two expansions must also be proportional to it.

The expansion of solids is not so often made use of for the measurement of temperature as that of fluids, both because it is much smaller in amount and is less easily measured.

80.   When a thermometer is brought into the neighbourhood of a medium or substance whose temperature is desired, its presence alters that temperature by the interchange of heat which immediately takes place, and which is indeed essential to the use of the thermometer; it is this new temperature which is the subject of observation, and not the original; under ordinary circumstances, however, when the mass of the thermometer is small compared with that of the substance around it (Art. 87), there is no appreciable difference between them.

81.   It is a curious circumstance with regard to water, that although it decreases in volume as its temperature diminishes down to about 40° Fahrenheit, for a still further diminution its volume increases again.* Some most important results following from this fact will be noticed below, Art. 111.

* Recent experiments seem to shew that this property belongs to many other substances as well as water.

82. It is discovered by experiment that the following very simple law regulates the expansion of gases under heat: —*the ratio of the increase of volume to the original volume is for all gases in the same proportion to the increase of temperature, provided the pressure exerted upon them remains unchanged, whatever that pressure be;* thus if V be the volume of a gas at a given pressure and temperature, V' its volume at the same pressure, but at a temperature increased by $t°$, then

$$\frac{V' - V}{V} = \alpha t,$$

where α is a number which is the same for all gases, but varies, of course, with the magnitude of the degrees in terms of which t is measured.

It appears then from this law and from Boyle's law (Art. 37) (each of them deduced from experiment) that the density of gas depends both on the temperature and pressure.

To find the relation between these three, suppose p_0, ρ_0, 0, p, ρ, t corresponding values of the pressure, density, and temperature for two conditions of a given gas. Then the change of density from ρ_0 to ρ would be produced, by first changing the temperature from 0 to t, and *then* changing the pressure from p_0 to p.

Let ρ_0 become ρ' by the first change, and let V_0, V_1 be the corresponding volumes of the gas. Then, by the law above enunciated, we have, since the pressure is unaltered,

$$\frac{V - V_0}{V_0} = \alpha t, \quad \therefore \frac{\frac{1}{\rho'} - \frac{1}{\rho_0}}{\frac{1}{\rho_0}} = \alpha t,$$

$$\text{or,} \quad \frac{1}{\rho'} = \frac{1}{\rho_0}(1 + \alpha t)$$

We have now a pressure p_0, a density ρ', and a temperature t. Now increase p_0 to p, the temperature remaining t. Then the density becomes ρ, and by Boyle's law (Art. 38),

$$\frac{p}{p_0} = \frac{\rho}{\rho'} = \frac{\rho}{\rho_0}(1 + \alpha t).$$

But whatever was the original density ρ_0, for a temperature 0, it was connected with the pressure by the relation

$$p_0 = k\rho_0, \quad (\text{Art. 38});$$

$$\therefore \ \frac{p}{k\rho_0} = \frac{\rho}{\rho_0}(1 + \alpha t),$$

$$\therefore \quad p = k\rho(1 + \alpha t),$$

the required relation.

It should be remembered carefully, that k and α are quantities which must be determined by experiment.

k is found to differ for different gases; α is the same for all. An explanation of a method for determining k for air has been given in Art. (38*).

83. We are unable to ascertain the amount of heat actually contained by a body when exhibiting a given temperature, but by careful thermometric experiments we can discover how much heat is absorbed or given out by it in passing from one known temperature to another; the amount of this heat so absorbed or evolved being estimated by the number of degrees to which it will raise the temperature of a given mass of water. The results of such experiments are, that for the same substance :

(1) The quantity of heat required to be absorbed or given out, in order to produce a *given* increase or decrease respectively of temperature, is proportional to the mass :

(2) The quantity absorbed or given out by a *given* mass is in a *constant* proportion to the consequent increase or decrease of temperature.

84. This may be concisely illustrated by saying that if m and m' be two masses of the same substance exhibiting temperatures t and t' respectively, and if τ be the uniform temperature which these two masses united will attain, then

$$(m + m')\tau = mt + m't'.$$

Thus if any volume of water heated to 70° of any thermometer, be mixed with twice the same volume of water at 100°, the mixture will be found to have the temperature of 90° by the same thermometer.

85. When we come to the consideration of different substances, we find the *constant* ratio of law (2) is different for each: if q represent the quantity of heat absorbed by a unit of mass of distilled water in order to increase its temperature by one degree of heat, and q' the quantity required for the same purpose by a unit of mass of mercury, q and q' are very different; of course their absolute magnitudes would, by both laws, vary in proportion to the magnitude of the unit of mass and the degree of heat; but their ratio, instead of being unity, is $\frac{q'}{q} = .033$ nearly: if q' refer to spermaceti oil, $\frac{q'}{q} = .5$.

This general fact may be got at by a variety of experiments. For instance, if equal weights of quicksilver and water be mixed, the first having a temperature 40°, and the second 156°, the temperature of the resulting mixture is found to be 152°.3; it thus appears that the water has lost *heat* while the mercury has gained some, and it cannot be doubted but that these quantities are equal; but then this quantity abstracted from the water only diminishes its temperature 3°.7, while it raises the temperature of the mass of mercury, to which it is added, and which is equal to the mass of water, 112°.3. From this it may be inferred, that for raising a given mass of mercury 1° it requires $\frac{3.7}{112.3}$ of the heat which will produce the same result in an equal mass of water.

Again, if a pound of water at 100° be mixed with two pounds of oil at 50°, the resulting temperature will be 75°; therefore the same heat which will lower 1 lb. of water 25°, will raise 2 lbs. of oil the same amount.

86. These quantities, q and q', are generally denominated the *specific heats* of the respective substances, and are most commonly measured in terms of the specific heat of some one substance, taken as the unit; if this one be distilled water, as is usually the case, we should have q', the specific heat of mercury, $= .033$; and similarly for all other substances.

87. To include these results under a general formula, let τ^0 be the temperature resulting from the combination of a mass m, say of mercury, at a temperature t^0, with a mass m', say of water, at a temperature $t^{0'}$; and let q, q', as before, be the specific heats of mercury and water respectively.

Suppose t to be greater than t', and C, estimated in terms of the same unit as q and q', the quantity of heat lost by the whole mass of mercury and therefore gained by the water; since then it would require a loss of heat $= mq$ to reduce the temperature of the mercury one degree, its actual reduction is $\dfrac{C^0}{mq}$; similarly the increase of the water's temperature must be $\dfrac{C^0}{m'q'}$; therefore we have

$$t - \frac{C}{mq} = t' + \frac{C}{m'q'} = \tau,$$

from which results, by the elimination of C,

$$(mq + m'q')\tau = mqt + m'q't' \dots\dots\dots (\alpha),$$

$$\therefore \frac{q'}{q} = \frac{m(t - \tau)}{m'(\tau - t')} \dots\dots\dots (\beta);$$

the equation (Art. 84) is evidently only a particular case of (α).

88. The foregoing definition of specific heat is not extended to gases. It is found more convenient in such cases to speak of two kinds of specific heat: the *first* refers to the quantity of heat required to raise by one degree the temperature of a gas contained by the rigid sides of vessels of *constant volume*: the *second*, to be the quantity required to increase by one degree the temperature of a gas contained by the extensible sides of a vessel of such nature that the *pressure* is maintained constant.

89. It has been remarked (Art. 76) that an alteration of the temperature is always accompanied by an alteration of the dimensions of a body, unless some constraining force be in action: it may be further asserted that all solid bodies may by a sufficient increase of heat be rendered liquid, and by a still greater increase, changed into vapour. The converse seems also to be true, that all elastic fluids will upon a withdrawal of heat become inelastic, and, if the process be continued long enough, at length solid. In short, alteration of temperature in any substance is always accompanied by either change of volume or change of chemical character. Also with elastic fluids increase or decrease of pressure produces the same effects as decrease or increase of temperature respectively.

90. The term *vapour* is usually applied to those elastic fluids which at ordinary atmospheric pressure and temperature lose their elasticity.   The laws connecting their changes of temperature, pressure, and fluid state, are very important, the more so as one of them, steam, is now of such extensive application as a motive power.

91. When an inelastic fluid occupies a portion of an inclosed vessel, the remainder being at first a perfect vacuum, a certain amount of vapour is disengaged from the fluid and passes into the empty space, until the space will hold no more at that temperature; this space is then said to be *saturated* with the vapour and the vapour itself to be at saturating density : the temperature at which a given density of vapour saturates space is termed the *Dew Point* of that density : the origin of this term will be explained in a later article (Art. 110). If the temperature be now increased, without any alteration in the volume of the inclosing vessel, the elastic force of the vapour increases also : and it does so to a much greater extent than accords with Boyle's law; this arises from the circumstance that additional vapour is generated by the fluid, and therefore the space in its new state of saturation corresponding to the new temperature contains more vapour than before. The law which connects the density of a vapour with its dew point is not simple, and need not be introduced here.

92. If the space containing the vapour be shut off from all communication with the fluid, an increase or decrease of temperature in the vapour will cause an alteration in its pressure, accordant with the law of Art. 82, down to the point when the temperature is just sufficient to retain the given vapour in an elastic state, *i.e.* down to the dew point of its density; at this point and below it, just so much of the vapour will be deposited in its liquid state as will leave the remainder at the required saturating density.

Again, if the temperature remaining constant, the space above the liquid be increased, sufficient vapour will be disengaged to keep the space saturated.

Generally, then, if vapour rise into a void space, it will, whether in contact with its own liquid or not, maintain a given density at a given temperature, a portion of it, if necessary, becoming condensed: excepting beyond the limiting temperature when its quantity is not sufficient to effect this, and then it follows Boyle's law.

93. There is every probability that what are usually termed permanent gases are vapours whose saturating densities are very great for low temperatures, and thence arises the difficulty of reducing them to the liquid state.

94. The results which have just been stated with respect to the quantity of vapour which rises at a given temperature into a void space from a liquid exposed to it, are also true when the space is already occupied by any other gases or vapours; the only difference is, that in this case time is required to complete the saturation.

95. When a space is occupied by any number of gases and vapours together (these last being supposed at saturation density, otherwise they are, as far as we are concerned, gases,) the laws of Art. (74) obtain for the mixture: it is uniform throughout, and the pressure at every point is the sum of those pressures which would be separately due to the individual gases or vapours if they occupied the space alone.

96. The curious phenomenon of ebullition which takes place when water is heated to boiling point admits of explanation by the aid of the foregoing principles. When the vessel is placed upon the fire the particles of water next its

sides become heated, rise to the surface, and there give off their vapour: the colder particles which take their places become heated in turn, and by a continuation of this process the whole mass becomes gradually heated; but so far the operation of heating is quite tranquil. At length the lower layers attain such a temperature, that the pressure of their vapour at the corresponding saturating density is greater than that of the superincumbent fluid: this vapour therefore expands, forms itself into bubbles and rises towards the surface; but in its progress it comes into fluid of a lower temperature and is consequently suddenly condensed; this produces the bubbling noise and disturbance in the water which precedes the boiling. These heated bubbles of vapour however greatly accelerate the equalization of the temperature of the whole fluid: it finally becomes uniform, and the bubbles of vapour generated in all parts of the fluid pass out to the surface at uniform pressure, which must manifestly equal that of the atmosphere: the disturbance arising from the passage of these bubbles continues, but the crackling and bubbling has ceased and the boiling is completed. Ebullition, then, which is the criterion of boiling, occurs as soon as the temperature of the interior particles of the fluid becomes such that the pressure of their vapour at the corresponding saturating density is greater than the pressure of the surrounding fluid.

We see from this explanation that the pressure of the atmosphere upon the surface of water must materially affect the temperature at which water will boil, and must therefore be taken into consideration in the graduation of the thermometer. The height of the barometer is about 30 inches when boiling water has the temperature 212° indicated by Fahrenheit's thermometer. (Art. 78.)

97. The preceding laws connecting the pressures, &c. of a mixture of a gas with vapour may be exhibited in an alge-

braical form. Suppose the gas to be always in contact with liquid affording the vapour, and its quantity to be given, then if V be the space it occupies at temperature t, and if p denote what would be its pressure if it occupied that space alone, by Art. (82),

$$p = \frac{k'}{V}(1 + \alpha t) \ldots\ldots\ldots (1),$$

because the quantity of gas remaining the same, its density varies inversely as its volume. Now by Art. (74), if P be the whole pressure of the mixture, F the elastic force or pressure of the vapour at saturating density,

$$P = p + F \ldots\ldots\ldots\ldots (2),$$

therefore, substituting above,

$$V = \frac{k'(1 + \alpha t)}{P - F} \ldots\ldots\ldots (3).$$

98. From the preceding remarks it may be collected that, when bodies change their state and dimensions, a portion of heat is always either absorbed or given out by them, and that this portion is not all accounted for by the consequent alteration in their temperature. In fact, one part of the heat absorbed or given out is employed in producing mechanical changes in the body itself, *i. e.* in making it larger or smaller, or in altering its chemical nature ; the other goes towards increasing or decreasing the temperature. The definition above given of *specific heat* takes the whole of this heat into consideration, while the first part is usually distinguished by the name of *latent* or *insensible heat.*

99. The quantity of heat which becomes latent in the passage of an inelastic fluid into the gaseous state is very great, all of which is absorbed by the vapour at the moment of its generation either from its parent fluid or from any solid

body with which it may then be in contact, such as would be the case if the evaporating liquid were percolating any substance. The vapour and liquid, like all bodies which are in contact, and which readily impart heat to each other, are always of the same *temperature*, and therefore during the evaporation, if no extraneous heat be supplied them, they will both gradually cool, and thus, if the pressure above the liquid be diminished sufficiently to keep it boiling notwithstanding the lowered temperature, a freezing of the liquid will actually be the result of its boiling.

With the aid of the air-pump many experiments may be made to shew this remarkable phenomenon: but perhaps the most instructive of all is the following: If two bulbs of glass be connected by a bent tube, and one be filled with water and then heated so that the vapour rises sufficiently to drive all air out of the tube through an orifice in the empty bowl, and if this orifice be then closed, the pressure of the included steam will after cooling be reduced to that which is due to the temperature of the air: if now the empty bulb be immersed in a freezing mixture, the steam will have its pressure so diminished that the water in the other bulb will immediately boil very rapidly, and the consequent vapour will carry off from it to the freezing mixture enough latent heat to convert the remaining water into ice. This apparatus was invented by Wollaston, and called by him the Cryophorus, or frost-bearer.

100. This explanation fully accounts for the well-known cooling effect of evaporation. A sudden dilatation of a gas produces the same result, as does also the reduction of a solid to a liquid state, such as the thawing of ice. On the contrary, physical changes of an opposite kind cause the body to evolve heat: thus when quicklime is slaked and a species of concretion or solidification of the water thereby produced, considerable heat is given forth. Also in some neat contrivances

for lighting matches the necessary heat is produced by a sudden forcible condensation of gas, or compression of a solid.

101. Many meteorological phenomena, such as clouds, fogs, rain, dew, &c., are caused by changes of temperature taking place in an atmosphere charged with moisture. We are now in a position to offer some explanation of them.

The water at the surface of the earth in contact with the atmosphere is continually giving off vapour, but not with sufficient rapidity, compared with the compensating causes, to produce general saturation: the portion of air nearest the earth's surface is found at mean temperature to contain about half the quantity of vapour required to saturate it. It may be here mentioned that the ratio which the quantity of vapour actually present at any time in a portion of air bears to the quantity which would saturate it at the temperature then existing, is taken as the measure of the *hygrometrical* state of the air at that time. The instruments, of which there are various kinds, used for ascertaining this ratio are termed hygrometers.

102. The vapour formed at the surface of the earth has a much less specific gravity than the air, it therefore rises rapidly, and becoming in consequence exposed to a diminution of atmospheric pressure, expands; this expansion cools the surrounding air, which was probably at a lower temperature already than that near the earth; and it thus happens that at a certain height the vapour is very nearly sufficient to produce saturation; here, therefore, if any reduction of temperature is by any means effected, vapour must be condensed. Now the first form of condensation of a mass of vapour is that which is exhibited by steam issuing into the atmosphere and which makes it visible: the vapour seems to condense into little *hollow* spheres, or vesicles, having a nucleus of air in

the center, which owing to the evolution of latent heat is of a temperature certainly not less than that which the vapour had previous to condensation; each sphere, together with its included air, possesses a specific gravity which need not be greater than that of the surrounding atmosphere, and thus the whole mass will float in a visible form, which we call a *cloud.*

103.   A still further diminution of temperature causes the vesicles of the clouds to collect into *solid* drops, which being necessarily of greater specific gravity than air, immediately fall to the surface of the earth and produce what is called *rain.*

104.   Changes of temperature of the kind here supposed may be often attributed to electrical causes, but generally, no doubt, they result from the passage of a current of hotter or colder air.   It is possible that masses of air, each charged with vapour, should by meeting form a mixture, whose reduced temperature would require for saturation less vapour than that which they all together brought with them; this would cause the instantaneous appearance of a cloud, and perhaps rain; and would very well account for a phenomenon which is by no means uncommon.

Any mechanical means which would bring a quantity of air already saturated into a space where a temperature lower than its dew-point obtains (Art. 110), would produce a manifestation of vapour or rain; it thus happens that the tops of mountains are very generally capped with clouds; for the currents of air charged with moisture, which are carried along by the winds nearly parallel to the surface of a level country, slide up the sides of any hills which they meet with, and are thus raised by them, as by inclined planes, into an elevation, where the temperature perhaps, for reasons mentioned in Art. 102, is always lower than it is below.

105. *Snow* results from the freezing of vapour at the moment of condensation, while *hail* proceeds from the freezing taking place after the drops have been formed, and during their passage, in falling, through a portion of air having a very low temperature.

106. *Fogs* and *mists* are only clouds in different states of density actually in contact with the earth's surface.

107. *Dew*, and *hoar frost*, and *night fogs* are phenomena of the same class as the preceding: although their immediately acting cause is peculiar. They are produced by the earth and the objects near its surface becoming colder than the superincumbent air,.and then acting upon it as a refrigerator: the layer next the earth may thus have its temperature so much reduced that the vapour contained by it is too much for its saturation, the superfluous quantity will then be deposited in the form of *dew ;* at the same time, a little higher up, the temperature may be only just low enough to exhibit the condensed vapour in the shape of a fog, while above this the air will be clear. If, again, the temperature of the earth under consideration be as low as freezing-point, the vapour will freeze upon condensation, and an effect similar to snow will be produced on the surface of the earth; this is *hoar* frost. It is sufficiently apparent that the air in valleys, and above streams and pieces of water, will receive more vapour in the course of the day than that elsewhere, and will therefore be the more nearly saturated; hence it is that these places are the most favourable for exhibiting fogs, dews, &c.

108. It remains now to explain how it is that the earth so often becomes colder at night than the air above, and thus produces the effects thus described. There are three distinct modes in which heat may be transmitted from body to body, *Conduction*, *Convection*, and *Radiation*. It is *conduction* when

the heat passes from one part of a body to another, through the intermediate particles, or from one body to another which is in *contact* with it: thus if one end of a metal wire be heated in a candle, the other will become hot by conduction; and both the warmth experienced when the finger is immersed in boiling-water, and the cold, when it touches ice, are the effects of conduction; in the first case it communicates heat from the water to the finger, and in the other it takes from the finger to give it to the ice.   Some bodies are so much better conductors than others, as to produce very different sensations of heat upon being touched by the hand, although the real temperature of all may be the same; all metals are so in comparison with wood, &c.; and of this fact practical use is made when ivory handles are attached to teapots.   Silk and wool are well-known non-conductors, and are therefore admirably adapted for clothes, to prevent the animal heat from escaping too rapidly.

When any fluid, elastic or not, after receiving heat, passes to another place, and there gives it up again, the process of transmission is termed *Convection*.

Radiation is of quite a different character; it is a term used to designate the process of interchange of heat which is continually going on between all substances at all distances from each other.   Each body seems to throw out its own heat in all directions around itself, just as light radiates from a luminous point, and to absorb that which comes to it in the same way from its neighbours.   It is an ascertained fact, that when a portion of the surface of any body receives heat from a source which is at the same temperature as itself, it absorbs exactly the same quantity which it radiates; and thus when equilibrium of heat is once established amongst bodies surrounding each other, it will be always maintained, unless disturbed by some extraneous cause: but if the portion of radiating surface be opposed to no source of heat whatever, it will lose all the heat which radiates from it, and the whole of the body of which it forms a part will cool by conduction.

109. Our promised explanation is now easy, for it so happens that the radiating power differs extremely with different substances, and that with air, as with all perfect fluids, it is almost zero; at the same time air presents little or no obstacle to the free radiation of heat through it: thus at night, when the sun and all extraneous sources of heat are removed, the surface of the earth and the bodies upon it will entirely lose just so much of their heat as radiates *through* the air into empty space, supposing no such objects as clouds, trees, &c., exist in that direction to return it; they may therefore cool sufficiently below the temperature of the air with which they are in contact, to effect by conduction the results described.

This explanation is confirmed by the extraordinary difference in the quantity of dew which is observed to be deposited in the same place, upon objects of different kinds; thus grass, and the leaves of most vegetables, glass, chalk, and generally bodies with rough or dense surfaces, all of which have a power of rapid radiation, will be covered with dew at a time when metals and many bodies with smooth surfaces lie almost dry by the side of them.

110. If a solid body be introduced any where into the atmosphere, and if its temperature be just below that point at which the vapour in the atmosphere around it is sufficient to saturate it, some of this vapour must be deposited upon its surface in the form of dew. Hence if the temperature of such a body be observed, as it is gradually made to cool down till dew is seen upon it, and if also it be observed when it is re-heated till this dew disappears again, the real temperature at which the vapour in that particular part of the atmosphere

saturates it, and which must evidently be intermediate to these
two, can be approximately ascertained.   This temperature is
called the *dew-point* (a term which we have already used) for
that portion of atmosphere at that time, and from it can be
obtained the actual quantity of vapour present, by the aid of
tables which connect the saturating density of vapour with the
corresponding temperature.

111.   The heating or cooling of liquids is almost entirely
effected by convection, as their conducting powers are slight.
This has been already assumed to be the case in the explanation given of the boiling of water, and is confirmed by the
circumstance, that to make water boil by heating it from
above is a process requiring much time: the upper particles
are in this case the first to have their temperature heightened,
and they do not then, as the lower would, under like circumstances, give place to others, because their diminished specific
gravity is a sufficient reason for their remaining where they
are.

When liquids cool from above, *i. e.* when they lose the
heat which radiates from their upper surfaces, each layer of
particles, as it becomes cooler than the rest, and therefore has
its specific gravity by the contraction of its volume increased
beyond that of the liquid below it, sinks, and is replaced by
the adjacent portions of the same liquid, which will in turn
undergo the same fate: thus the whole mass has its temperature uniformly reduced; and if this reduction goes as far
as the freezing-point, the whole will be congealed.

If such were the result of the cooling of water, it would
be most disastrous in the present condition of our globe: all
aquatic life would be destroyed every time that a severe frost
occurred, and a lake or river when once converted into solid
ice would never melt again in our climates, for the heated

water resulting from the partial thawing of the surface, would not, for the reasons just given, convey downwards sufficient heat for the thawing of the remainder. Fortunately for us the law given above, by which water contracts, prevents these unpleasant consequences: after a volume of water has been reduced down to a uniform temperature of 40°, upon a continuation of the cooling process, its upper particles will cease to descend, and will soon become a sheet of ice, which by its non-conducting power materially preserves the remaining water from the further effect of cold.

GENERAL EXAMPLES.

(1) A thin conical surface (weight W) just sinks to the surface of a fluid when immersed with its open end downwards: but when immersed with its vertex downwards, a weight equal to mW must be placed within it to make it sink to the same depth as before: shew that, if a be the length of the axis, h the height of the column of the fluid, the weight of which equals the atmospheric pressure,

$$\frac{a}{h} = m\sqrt[3]{1+m}.$$

It is evident from the question that the weight of the fluid displaced in the second immersion is to the weight of that displaced in the first $:: 1+m : m$.

But the volume displaced in the first case is that occupied by the compressed air within the shell, and is therefore a cone whose axis may be represented by z, while that in the second case is the volume of the whole cone; now the volumes of these cones are in the ratio of the cubes of their axes; therefore,

$$\frac{a^3}{z^3} = 1 + m \dots\dots (1).$$

Again, the pressure of the air compressed into the cone whose axis is z must balance the pressure due to the depth z below the surface of the fluid; and before compression when it occupied the whole volume of the cone, its pressure balanced a column of fluid equal to h; therefore

$$\frac{a^3}{z^3} = \frac{z+h}{h} \quad \dots\dots\dots\dots (2),$$

hence, substituting in (1),

$$\frac{z+h}{h} = 1 + m,$$

or $z = mh$,

and therefore again from (1) $\frac{a}{h} = m\sqrt[3]{1+m}$.

(2) A hollow conical vessel floats in water with its vertex downwards and its base on the level of the water's surface: it is retained in that position by means of a cord, one end of which is attached to the vertex and the other to the center of a circular disc lying in contact with the horizontal plane upon which the water rests; given the dimensions of the cone and the depth of the water, find the smallest disc which will answer the purpose, neglecting the weight of the cord, cone, and disc.

As there is no fluid *under* the disc the resultant fluid pressure upon it is the same as the total pressure upon its upper surface: it is therefore equal to the weight of a cylindrical column of the fluid having the disc for its base and the depth of the fluid for its height: since the horizontal plane can only exert a force of resistance, the smallest disc is evidently that, upon which this downward fluid pressure is just sufficient to balance the tension of the string upwards, when the cone is in the given state of immersion: but the force required to be exerted by the string in order to hold the cone in this position is equal and opposite to the resultant of the fluid pressures upon the surface of the cone, *i.e.* to the weight of the fluid displaced by it. Hence we conclude that the smallest disc required will equal the base of a cylinder of the fluid, whose altitude is that of the fluid and whose volume is that of the given cone.

(3)   Find the depth in water at which the pressure is 140 lbs., assuming the atmospheric pressure to be 15 lbs. the square inch, and an inch the unit of length.

(4)   A vertical cylinder contains four cubic feet of water, of depth nine inches; find the pressure in lbs. at any point in the base, considering four inches the unit of length, and assuming one cubic foot of water to weigh 1000 ozs.

(5)   A body in the form of an equilateral triangle floats in water; determine the condition to be satisfied in order that one angular point may be in the surface of the water and the opposite side vertical.

The centers of gravity of both the triangle and the fluid displaced by it may be easily proved to be in the same vertical line; the condition referred to therefore concerns the specific gravities of the triangle and fluid.

(6)   A pyramid with a square base and with sides which are equilateral triangles is placed on a horizontal plane and filled with a fluid through an aperture in the vertex; find the pressure on one of the sides.

If the pyramid have no base, find its least weight consistent with its not being raised from the plane.

(7)   The surface of a man's body contains $14\frac{1}{2}$ square feet; find the pressure on it when at a depth of 20 fathoms in salt water whose specific gravity is 1.026.   State also how the resultant of this pressure might be found.

(8)   If a cubic inch of distilled water weighs 253 grains, and the specific gravity of salt water be 1.026, what will be

the pressure on a square inch at a depth of 20 feet below the level of the sea?

(9)   The lighter of two fluids (s. g. 1 : 2) rests to a depth of four inches on the heavier, a square is immersed vertically with one side in the surface; determine the side of the square that the pressure on the portions in the two fluids may be equal.

(10)   A hemispherical vessel with its base horizontal is filled with fluid through an orifice at its highest point; prove that the whole pressure on the curved surface equals that on the base.

(11)   A cylindrical vessel with its axis vertical is filled with equal masses of two fluids which do not mix; compare their densities, supposing the pressures on the upper and lower portions of the concave surfaces equal.

(12)   A cone with its axis vertical and base downwards is filled with fluid; find the normal pressure on the curved surface, and compare it with the weight of the fluid.

(13)   A cubical box filled with a fluid of a given weight W, is supported in such a position that one of its edges is horizontal, and that one of its sides passing through this edge is inclined at an angle α to the horizon; shew that the sum of the pressures on the six faces is equal to

$$3\,W\,(\sin\alpha + \cos\alpha).$$

(14)   An isosceles triangle has its vertex in the surface of a fluid and base parallel to it; find the pressure and center of pressure.

(15)   A figure bounded by the arc of a parabola AP, the tangent at the vertex AB, and the line PB parallel to the axis, is immersed vertically in a uniform fluid, with A in the surface and BP horizontal; find the depth of the center of pressure.

(16) If the side of a rectangle be horizontal and at a given depth below the surface of a fluid, determine the whole pressure on the rectangle; and shew that the center of pressure lies below the center of gravity of the rectangle.

(17) A hemispherical bowl is filled with fluid, and different sections of it are taken through the same tangent line to its rim; determine the section upon which the pressure is the greatest.

(18) A portion of a paraboloid, of density ρ, cut off by a plane perpendicular to its axis, floats with its axis vertical in a cylinder containing two fluids, of densities ρ and 8ρ, which do not mix. Having given that the radius of the cylinder, the latus rectum, and length of axis of the paraboloid are all equal, find the volume of the upper fluid when the two ends of the axis of the paraboloid project equal distances above and below the surface of that fluid when there is equilibrium.

(19) ABC is a right-angled triangular plate, and it floats with its plane vertical and the right angle C immersed in water; prove that if its specific gravity be to that of water as $2:5$, and $CB:CA = 5:4$, CB is cut by the surface of the water at a distance from $C = CA$.

(20) One end of a uniform rod is attached to a hinge fixed in a mass of fluid; to the other is attached by a free joint the vertex of a cone which floats in the fluid. Given that the volume of the cone is 9 times that of the rod and the specific gravity of the rod 20 times that of the cone, find the specific gravity of the fluid in order that the cone may float with $\frac{2}{3}$ of its axis immersed.

(21) A right cylinder of radius a and height $2h$ floats in a fluid of double its density with one of its circular ends

entirely out of the fluid; shew that it can rest with its axis inclined at a certain angle to the vertical if $h > \dfrac{a}{\sqrt{2}}$.

(22) A piece of zinc (whose specific gravity is 6.9) weighs 59 ozs. in distilled water and 61 in alcohol; find the specific gravity of alcohol.

(23) A lump of metal weighs 59 ozs. in water and 61 ozs. in alcohol whose specific gravity is ·8; find its weight and specific gravity.

(24) A uniform cylinder when floating with its axis vertical in distilled water sinks to a depth of 3.2 inches, and when floating in alcohol sinks to a depth of 4 inches; find the specific gravity of alcohol.

(25) A vessel, of weight W times that of the cubic foot of water, in sailing down a river leaks V cubic feet of water, and is observed to be immersed to a given depth. On reaching the sea V' cubic feet are pumped out; and after V'' cubic feet of sea-water have been leaked, the vessel is observed to be immersed at the same depth as before: find the specific gravity of sea-water.

Obtain a numerical result, taking specific gravity of fresh water $= 1$, weight of ship $= 100$ tons, $V = 1000$, $V' = 500$, $V'' = 600$.

(26) Find the greatest amount of water displaced by the air in a cylindrical diving-bell.

Also find the interior pressure then on the upper surface of the bell.

(27) Given h_1, h_2, the heights of the barometer in a diving-bell before descent and at a certain depth respectively, find the depth.

(28)   Assuming that 100 cubic inches of air weigh 31.0117 grains and a cubic foot of water weighs 1000 ozs., compare the specific gravities of air and water: and if 34 feet be the height of a column of water which the atmosphere will support, shew that the height of the atmosphere considered homogeneous is about 5 miles.

(29)   The weight of a cubic foot of water being 1000 ozs. and its specific gravity unity, determine the specific gravity of a substance whose bulk is m cubic inches and weight n ozs.

(30)   A cubic foot of water weighs 1000 ozs.; what is the specific gravity of a solid of which a cubic yard weighs 540 lbs.?

(31)   A cube of wood floating in water descends 1 inch when a weight of 30 ozs. is placed on it; find the size of the cube, supposing a cubic foot of water to weigh 1000 ozs.

(32)   If the specific gravity of air be s, that of water being 1, and if W, W' be the weights of a body in air and water respectively, shew that its weight in vacuo will be

$$W + \frac{s}{1-s}(W - W').$$

(33)   A metal cylinder floats in mercury with one-fourth of its bulk above the surface; find the specific gravity of the metal, that of mercury being 13.6.

(34)   A piece of wood weighs 6 lbs. in air; a piece of lead which weighs 12 lbs. in water is fastened to it, and the two together weigh 10 lbs. in water; find the specific gravity of the wood.

(35)   What weight of oil (specific gravity .75) must be added to a pound of fluid of specific gravity .5, that in the

mixture a pound of a substance of specific gravity 4 may weigh 13.5 ozs.?

(36) To a piece of wood which weighs 4 ozs. in vacuo a piece of metal is attached whose weight in water is 3 ozs. and the two together are found to weigh 2 ozs. in water; find the specific gravity of the wood.

(37) A lump of silver weighs 550 grains in air and 506 grains in water, find the specific gravity of silver, and also the volume of the lump, having given that the weight of a cubic inch of water is 250 grains.

(38) A cubical block of marble whose edge measures 2 feet and whose specific gravity is 2.7 has to be raised out of a river; determine its weight when entirely immersed, and also when lifted out of the water.

(39) Two bodies A and B in air weigh 10 lbs. and 15 lbs. respectively; in mercury B alone and A and B together weigh respectively 9 lbs. and 1 lb.: what is A's specific gravity, that of mercury being 13.5?

(40) A cubic inch of pure gold (specific gravity $16\frac{1}{2}$) is mixed with two cubic inches of mercury (specific gravity 13.6); find the specific gravity of the compound.

(41) What weight of water must be added to a pound of fluid whose specific gravity is $\frac{1}{2}$ in order that the specific gravity of the mixture may be $\frac{3}{4}$?

(42) The apparent weight of a sinker, in water, is four times the weight in vacuum of a piece of material, whose specific gravity is required: that of the sinker and piece together is three times the weight. Shew that the specific of the material $= .5$.

(43) If the three weights used in Nicholson's Hydrometer be 10, 12, and 18 lbs., find the volume of the solid in inches; a cubic foot of the fluid weighing 1000 ozs.

(43*) If p parts by weight of a metal whose specific gravity is s, when fused with p' parts of a metal whose specific gravity is s', form an alloy whose specific gravity is S, shew that $\dfrac{1^{\text{th}}}{n}$ part of the volume of the whole has been lost by condensation during the mixture where

$$\frac{1}{n} = 1 - \frac{ss'(p+p')}{S(ps'+p's)}.$$

(44) Find k, for air, that $p = k\rho$ may give the pressure in ounces; the barometer standing at 30 inches when the density of air referred to mercury is .0001; the unit of length being one inch, and a cubic foot of the standard substance weighing 1000 ozs.

How will k be altered if the unit of length be increased to 6 inches?

(45) p_1, ρ_1, t_1, p_2, ρ_2, t_2, p_3, ρ_3, t_3, are corresponding values of the pressure, density, and temperature of the same gas; shew that

$$t_1\left(\frac{p_2}{\rho_2} - \frac{p_3}{\rho_3}\right) + t_2\left(\frac{p_3}{\rho_3} - \frac{p_1}{\rho_1}\right) + t_3\left(\frac{p_1}{\rho_1} - \frac{p_2}{\rho_2}\right) = 0.$$

(46) The temperature at one place is 24° by the centigrade and at another 52° by Fahrenheit; what is the difference by Fahrenheit's?

(47) What is meant by the sensibility of the thermometer? What degree of a centigrade corresponds to 60 of Fahrenheit, and what degree of Fahrenheit's to 60 of the centigrade?

(48) Having given a certain temperature in degrees according to Fahrenheit's thermometer, find the number of de-

grees indicating it on De Lisle's thermometer, where the space between boiling and freezing point is divided into 150 degrees, and the boiling point is taken as the zero of the scale.

(49) The point at which mercury freezes is indicated by the same number on the centigrade and on Fahrenheit's scale: determine the number.

(50) In a vessel not quite full of water, and closed at the top by a flexible membrane, a small glass balloon, open at the lower part, contains sufficient air just to make it float: explain the principle upon which the balloon sinks when the membrane is pushed in.

(51) A weightless conical shell is filled with fluid and suspended by its vertex from a fixed point: it is then divided symmetrically by a vertical plane, and kept from falling asunder by a hinge at the vertex, and a ligament at the base, coinciding with that diameter of the base which is perpendicular to the dividing plane: determine the tension of the ligament.

(52) A closed vessel is filled with water containing in it a piece of cork which is free to move: if the vessel be suddenly moved forward by a blow, shew that the cork will shoot forward relatively to the water.

(53) A piece of cork is attached by a string to the bottom of a bucket of water so as to be completely immersed, and the bucket being placed in the scale of a balance is supported by a weight in the other scale; if the string be cut, will the weight begin to rise or fall? State your reasons.

(54) A cylindrical vessel containing fluid revolves uniformly about its axis with an angular velocity ω, and a solid cylinder of less specific gravity than that of the fluid floats in

it with its axis coincident with that of the revolving vessel;
find how deep it is immersed.

(55)   A transparent closed cylinder filled with fluid, in
which there are extraneous particles, some lying at the bottom
and some floating at the top, is set revolving about its axis:
it is then observed that the floating particles all flow in
towards the axis, while those at the bottom recede from it:
explain this.

(56)   If a vertical cylinder containing heavy fluid revolves
about a generating line with a uniform angular velocity, the
depth to which the surface sinks below its original level : the
height to which it rises above that level :: 3 : 5.

(57)   A circular tube is half full of fluid, and is made to
revolve uniformly round a vertical tangent-line with angular
velocity ω : if a be the radius, prove that the diameter passing
through the open surfaces of the fluid is inclined at an angle
$\tan^{-1} \dfrac{\omega^2 a}{g}$ to the horizon.

(58)   If a spherical envelope, of thickness k and radius r,
be formed of a substance, which, if made into a line having
a section K, would bear a weight W: find the number of
strokes of the piston after which this envelope placed under
the receiver of an air-pump would burst.

(59)   A vertical cylindrical vessel, closed at the base, is
formed of staves held together by two strings, which serve as
hoops, and is filled with fluid; shew that the tension of the
upper string is to that of the lower :: $h - 3a : 2h - 3a$, where
h is the altitude of the cylinder, and a the distance of the
upper and lower strings from the top and bottom of the
cylinder respectively.

(60) In the case of the previous question, how much of the fluid must be withdrawn from the cylinder in order that the tension of the upper string may vanish?

(61) A cylindrical boiler, the interior radius of which is 10 inches, and the thickness $\frac{1}{10}$ of an inch, is formed of a material such that a bar of it, one square inch in section, can just support a weight of 10,000 lbs. without being torn asunder; find the greatest pressure which the boiler can sustain without bursting.

(62) At 18°.9 (centigrade) the weight of a cubic foot of distilled water is 997.84 ozs., and at 16°$\frac{2}{3}$ its weight is 998.24 ozs.; find the temperature at which it shall be 1000.

(63) A cubic inch of water which weighs 252.458 grains will produce a cubic foot of steam *at atmospheric pressure;* find the specific gravity of steam.

(64) A quantity of air under the pressure of m lbs. to the square inch, occupies n cubic inches when the temperature is $t°$; find its volume under a pressure of m' lbs. to the square inch when the temperature is $t'°$.

(65) Having given that m lbs. of steam at the boiling-point, mixed with n lbs. of water at temperature t, produces $m + n$ lbs. of water at the boiling-point, compare the latent heat of steam and the specific heat of water.

(66) Upon what principle might the height of a mountain be approximately found by observing the temperature at which water boils at the top?

(67) A vessel contains air at atmospheric pressure; find the force in pounds necessary to be applied to a piston of

area A in the vessel to prevent its being forced out when the air is heated to temperature T above what it was at first.

(68)   A thermometer-tube, open at the top and filled with mercury, contains 1000 grains at $32°$ temperature; if the tube be heated till its temperature is $84°$, find how many grains of mercury will be expelled. The expansion of mercury in volume between $32°$ and $212°$ being .018, and the linear expansion of glass between the same points .0008.

ANSWERS TO EXAMPLES.

SECTION I.

(1) 2 feet nearly.

(2) 2.5 seconds.

(3) .25 of a second.

(4) .016.

SECTION II.

(1) $P = \dfrac{a}{A+a} \cdot W.$

(5) $1:2:4:5.$

(6) $35\frac{4}{23}$ lbs.

(7) 1.2 of a foot.

(8) Half the weight of the fluid.   No, the pressure on the table equals the whole weight of the fluid and vessel.

(9) $3:1$; the same in both cases.

(10) $\frac{1}{2}$ pound.

(12) 156 lbs. $\frac{1}{4}$ oz.

(13) It rises $\dfrac{1}{3073}$ of its height nearly.

(14) 2 nearly.

(19) $1:5.$

(20) $1:8.$

(21) 11.48 ozs. nearly.

(23) Ratio of volumes $= 19:2$; ratio of weights $= 17:11.$

(24) 1 lb.

(25) .98.

(26) .39.

(27) 48 inches; 10.5408; $236\frac{1}{9}.$

(29) $250:147.$

(32) $23:37.$

(34) When high.

(35) $60°.$

SECTION III.

(7) 60 lbs.

(9) 5 inches.

(10) 3 lbs. nearly.

(11) $\frac{1}{4}$ of height from base.

(15) $1:1083077.$

(16) 26000 feet.

(17) $20\frac{4}{7}$ inches.

(18) $\frac{1}{2}$ that of the atmosphere.

(20) 10.

(21) 35 ft. nearly.

(22) When AB, BC are equally inclined to the vertical.

(23) 28 ft. 3½ in.   (24) $\dfrac{\sigma_1 - \sigma_2}{2\sigma_1} \cdot V$.   (25) Increase.

(26) The depth of the surface of the water in the bell $= h' \cdot \dfrac{a - h}{h}$ where h' is the height of the water-barometer, h the altitude of the cylinder of water of the same weight and transverse section as the bell, and a the altitude of the bell.

(27) $\dfrac{h + d}{h}$ times the original quantity, where h is the height of the water-barometer, and d the depth of the lower rim of the bell.

(28) Their apparent weights in the given positions must be equal. In the second case the water must be at the same level in both.

(29) All the air would rise in bubbles, and the bell would sink.

(30) 2528 tons nearly.

SECTION IV.

(5) 15 strokes.

(6) The string must pass round in a horizontal plane at one-third of the height of the prism from the base.

GENERAL EXAMPLES.

(3) 288 feet.   (4) 7¾ lbs.

(5) The specific gravities must be in the ratio of 1 : 2.

(6) Pressure $= \dfrac{\sqrt{3}}{2} \times$ weight of fluid. The least weight of the pyramid $= 2$. weight of fluid.

(7) 49 tons 16 cwt. 25 lbs. 8 ozs. + pressure of atmosphere. The resultant pressure $=$ the weight of water displaced.

(8) 62298.72 grains.   (9) $2(\sqrt{3}+1)$ inches.

(11) $\rho_1 : \rho_2 :: 1 : 3$.

(12) Normal pressure $= \dfrac{2}{\sin \alpha} \cdot W$ where α is the semi-vertical angle of the cone, and W is the weight of the fluid.

(14) For the pressure, see Art. 18. The center of pressure is on the bisecting line at $\frac{3}{4}$ of its length from the vertex.

(15) Four-fifths of AB.

(16) For the pressure, see Art. 18.

(17) Inclination to horizon $= 30^0$.

(18) $\frac{1}{2}$ the paraboloid.   (20) Six times that of the cone.

(22) .8.   (23) 69 ozs.; 6.9.   (24) .8.

(25) $\dfrac{W + V}{W + V - V' + V''}$.

(26) The amount displaced when the upper surface is level with the fluid, equals the weight of water displaced.

(27) $\sigma(h_2 - h_1)$ where σ is the specific gravity of mercury referred to water.

(29) $1.728 \times \dfrac{n}{m}$.   (30) .32.   (31) The edge $= 7.2$ inches.

(33) 10.2.   (34) .75.   (35) $1\frac{1}{2}$ lb.   (36) .8.

(37) 12.25; .176 inches.   (38) 850 lbs.; 1350 lbs.

(39) 7.5.   (40) 14.56.   (41) 2 lbs.   (43) .032 feet.

(44) 173611.$\dot{1}$; 36 fold.   (46) 23.2⁰.   (47) $15\frac{5}{9}$; 140.

(48) $\dfrac{1060 - F}{6}$ where F denotes the number of degrees Fahrenheit.   (49) -40.

(50) The volume of air between the membrane and water is diminished, and therefore its pressure is increased. Hence the pressure throughout the fluid is increased, and therefore the volume of air in the glass balloon is diminished. Therefore the specific gravity of the balloon and included air, considered as one body, is increased, and, being at first just equal to, is now greater than that of the water, and the balloon consequently sinks.

(51) $\dfrac{3 - \tan^2\alpha}{2\pi \cdot \tan\alpha} \cdot W$ where W is the weight of the fluid, and α is the half of the vertical angle of the cone.

(53) To descend: for, when the string is cut, the cork rises, and some heavier fluid takes its place; the center of gravity of the bucket and its contents descends; *less* force therefore is called into action at that end of the balance than was the case when this center of gravity was in equilibrium; hence the weight at the other end of the balance is no longer supported, and consequently begins to descend.

(54) $\dfrac{\sigma'}{\sigma} h + \dfrac{r_1}{4g} \cdot \omega^2 \cdot r_1$, where r_1 is the radius, h is the height of the floating cylinder, and $\sigma\sigma'$ are the specific gravities of the fluid and cylinder respectively.

(55) Let v be the volume of one of the particles, ρ its density, ρ' the density of the fluid, ω the angular velocity of the cylinder, and r the distance of the particle from the axis. Then the particle will move towards the axis or recede from it, according as the resultant force acting on it towards the axis be greater or less than $\rho v \cdot \omega^2 r$. But this force equals $\rho' r \omega^2 r$ (see Art. 67);

therefore the particle will tend to the axis or away from it, according as $\rho' >$ or $< \rho$;

i. e. the lighter particles will flow towards, whilst the heavier will recede from, the axis.

(58) The number of strokes is the value of n given by the inequality

$$\Pi\left\{1 - \left(\frac{A}{A+B}\right)^n\right\} \frac{r}{2} \text{ first greater than } \frac{k}{K} \cdot W,$$

where A and B are respectively the volumes of the receiver and piston-barrel.

(60) To within a distance $3a$ of the bottom.

(61) 100 lbs. per square inch.

(62) 6.84 degrees centigrade.

(63) $\dfrac{1}{1728}$ referred to water.

(64) $\dfrac{m}{m'}\{1 + a(t - t')\}n$ cubic inches nearly.

(65) $\dfrac{\text{latent heat of steam}}{\text{specific heat of water}} = \dfrac{n}{m}(212 - t)$ where t is in degrees Fahrenheit.

(66) On the principle that the temperature at which water boils depends upon the atmospheric pressure, and that the one being given, the other is known. Thus the observed boiling temperature would give the height of the barometer, and thence the height of the mountain (Art. 40).

(67) $aT.\Pi A$, where Π is the pressure of the atmosphere in lbs. on the unit of area, at the temperature from which T is measured.

(68) 5.19 grains nearly.

THE END.

CAMBRIDGE:—PRINTED BY JONATHAN PALMER.

MATHEMATICAL BOOKS

PUBLISHED BY

MACMILLAN AND CO.

....

AN ELEMENTARY TREATISE ON DIFFERENTIAL
EQUATIONS. By GEORGE BOOLE, D.C.L. F.R.S., Professor of
Mathematics in the Queen's University, Ireland. A New Edition, revised
by J. TODHUNTER, M.A., F.R.S. Crown 8vo., cloth, 14s.

A SUPPLEMENTARY VOLUME, Crown 8vo., cloth,
price 8s. 6d.

A COLLECTION OF ELEMENTARY TEST QUES-
TIONS IN PURE AND MIXED MATHEMATICS; with Answers.
And Appendices on Synthetic Division, and on the Solution of Numerical
Equations by Horner's Method. By JAMES R. CHRISTIE, F.R.S.,
F.R.A.S., late First Mathematical Master at the Royal Military Academy,
Woolwich. Crown 8vo., cloth, price 8s. 6d.

ARITHMETICAL EXAMPLES PROGRESSIVELY
ARRANGED; together with Miscellaneous Exercises and Examination
Papers. By the Rev. T. DALTON, M.A., Assistant-Master at Eton
College. 18mo., cloth, price 2s. 6d.

A TREATISE ON THE CALCULUS OF FINITE
DIFFERENCES. By GEORGE BOOLE, D.C.L., Professor of Mathematics in the Queen's University, Ireland. Crown 8vo., cloth, 10s. 6d.

TAIT AND STEELE. A TREATISE ON THE
DYNAMICS OF A PARTICLE, with numerous Examples. By PETER
GUTHRIE TAIT, M.A., late Fellow of St. Peter's College, Professor of
Natural Philosophy in the University of Edinburgh; and the late WM.
JOHN STEELE, B.A., Fellow of St. Peter's College. *Second Edition.*
Price 10s. 6d.

THE ELEMENTS OF PLANE AND SPHERICAL
TRIGONOMETRY. By J. C. SNOWBALL, M.A., Fellow of St. John's
College, Cambridge. *Tenth Edition.* Crown 8vo., cloth, 7s. 6d.

AN ELEMENTARY TREATISE ON PLANE TRI-
GONOMETRY. With a numerous Collection of Examples. By R. D.
BEASLEY, M.A., Fellow of St. John's College, Cambridge, Head-Master
of Grantham Grammar-School. *Second Edition.* Crown 8vo., cloth, 3s. 6d.

A TREATISE ON ELEMENTARY MECHANICS.

For the Use of the Junior Classes at the University and the Higher Classes in Schools. With a Collection of Examples. By S. PARKINSON, B.D., President of St. John's College, Cambridge. *Third Edition*, revised. Crown 8vo., cloth, 9s. 6d.

A TREATISE ON OPTICS. By S. PARKINSON, B.D.

Crown 8vo., cloth, 10s. 6d.

A GEOMETRICAL TREATISE on CONIC SECTIONS.

With Copious Examples from the Cambridge Senate-House Papers. By W. H. DREW, M.A., of St. John's College, Cambridge, Second Master of Blackheath Proprietary School. *Third Edition.* Crown 8vo., cloth, 4s. 6d.

SOLUTIONS TO PROBLEMS CONTAINED IN MR.

DREW'S TREATISE ON CONIC SECTIONS. Crown 8vo., cloth, 7s. 6d.

GEOMETRICAL CONICS, including Anharmonic Ratio

and Projection. With numerous Examples. By C. TAYLOR, B.A., Scholar of St. John's College, Cambridge. Crown 8vo., cloth, 7s. 6d.

AN ELEMENTARY TREATISE on CONIC SECTIONS

AND ALGEBRAIC GEOMETRY. With a numerous Collection of Easy Examples progressively arranged, especially designed for the use of Schools and Beginners. By G. HALE PUCKLE, M.A., Principal of Windermere College. *Second Edition*, enlarged and improved. Crown 8vo., cloth, 7s. 6d.

ALGEBRAICAL EXERCISES. Progressively arranged

by Rev. C. A. JONES, M.A., and C. H. CHEYNE, M.A., Mathematical Masters in Westminster School. Post 8vo., cloth, price 2s. 6d.

Works by **ISAAC TODHUNTER, M.A., F.R.S.**

EUCLID FOR COLLEGES AND SCHOOLS. *Second*
Edition. 18mo., bound in cloth, 3s. 6d.

ALGEBRA FOR BEGINNERS. With numerous Ex-
amples. 18mo., bound in cloth, 2s. 6d.

A TREATISE on the DIFFERENTIAL CALCULUS.
With numerous Examples. *Fourth Edition.* Crown 8vo., cloth, 10s. 6d.

A TREATISE ON THE INTEGRAL CALCULUS.
Second Edition. With numerous Examples. Crown 8vo., cloth, 10s. 6d.

A TREATISE ON THE ANALYTICAL STATICS.
With numerous Examples. *Second Edition.* Crown 8vo., cloth, 10s. 6d.

A TREATISE ON THE CONIC SECTIONS. With
numerous Examples. *Third Edition.* Crown 8vo., cloth, 7s. 6d.

ALGEBRA for the USE of COLLEGES and SCHOOLS.
Third Edition. Crown 8vo., cloth, 7s. 6d.

PLANE TRIGONOMETRY FOR COLLEGES AND
SCHOOLS. *Third Edition.* Crown 8vo., cloth, 5s.

A TREATISE ON SPHERICAL TRIGONOMETRY
FOR THE USE OF COLLEGES AND SCHOOLS. *Second Edition.*
Crown 8vo., cloth, 4s. 6d.

EXAMPLES of ANALYTICAL GEOMETRY of THREE
DIMENSIONS. Crown 8vo., cloth, 4s.

Works by BARNARD SMITH, M.A.

A SHILLING BOOK of ARITHMETIC for NATIONAL AND ELEMENTARY SCHOOLS. 18mo., cloth.

The Shilling Book of Arithmetic is also published in Parts, to meet the convenience of very Elementary Classes, and will be sold as follows:—

PART I., containing the First Four Rules, in 32 pages, sewed in neat paper covers, price 2d. This Part contains all that is required of Standards I., II., and III., in the Government Examination.

PART II., containing the Compound Rules, Bills of Parcels, and Practice, in 48 pages, sewed in neat paper covers, price 3d. This Part contains all that is required of Standards IV., V., and VI., in the Government Examination.

PART III., containing Fractions, Decimals, Rule of Three, The Metric System, &c., in 112 pages, sewed in neat paper covers, price 7d.

The Three Parts, complete in 1 Vol., with the Answers, 18mo., cloth, price 1s. 6d.

ARITHMETIC AND ALGEBRA in their PRINCIPLES AND APPLICATION: with numerous Systematically-arranged Examples taken from the Cambridge Examination Papers. *Ninth Edition.* Crown 8vo., cloth, 10s. 6d.

ARITHMETIC FOR THE USE OF SCHOOLS. *New Edition.* Crown 8vo., cloth, 4s. 6d.

A KEY TO THE ARITHMETIC FOR SCHOOLS. *Fourth Edition.* Crown 8vo., cloth, 8s. 6d.

EXERCISES IN ARITHMETIC. Crown 8vo., 2s.; or, with Answers, 2s. 6d. Also, sold separately, in two Parts, 1s. each; Answers, 6d.

THE SCHOOL CLASS-BOOK OF ARITHMETIC. Parts I. and II., 18mo., limp cloth, price 10d. each. Part III. 1s.; or Three Parts in one Vol., price 3s. 18mo., cloth, (forming one of Macmillan's Elementary School Class-Books.

KEY TO CLASS-BOOK OF ARITHMETIC. Complete, 18mo., cloth, price 6s. 6d.; or, separately, Parts I., II., and III., 2s. 6d. each.

KEY TO A SHILLLING BOOK OF ARITHMETIC. Fcap. 8vo., cloth, price 4s. 6d.

A CATALOGUE

OF

EDUCATIONAL BOOKS,

PUBLISHED BY

MACMILLAN AND CO.,

BEDFORD STREET, STRAND, LONDON.

MACMILLAN'S CLASSICAL SERIES, for COLLEGES and SCHOOLS, being select portions of Greek and Latin authors, edited, with Introductions and Notes at the end, by eminent scholars. The series is designed to supply first rate text-books for the higher forms of Schools, having in view also the needs of Candidates for public examinations at the Universities and elsewhere. With this object the editors have endeavoured to make the books as complete as possible, passing over no difficulties in the text, whether of construction or of allusion, and adding such information on points of Grammar and Philology as will lead students on in the paths of sound scholarship. Due attention moreover is paid to the different authors, in their relation to literature, and as throwing light upon ancient history, with the view of encouraging not only an accurate examination of the letter, but also a liberal and intelligent study of the spirit of the masters of Classical Literature.

The books are clearly printed in fcap. 8vo., and uniformly bound in neat red cloth.

The following volumes are ready :—

CICERO—THE SECOND PHILIPPIC ORATION. From the German of Karl Halm. Edited, with Corrections and Additions, by JOHN E. B. MAYOR, Professor of Latin in the University of Cambridge, and Fellow and Classical Lecturer at St. John's College. New edition, revised. 5*s*.

10,000.1.79.

THE CATILINE ORATIONS. From the German of Karl Halm. Edited, with Additions, by A. S. WILKINS, M.A., Professor of Latin at the Owens College, Manchester. New edition. 3s. 6d.

THE ACADEMICA. Edited by JAMES REID, M.A., Fellow of Caius College, Cambridge. 4s. 6d.

DEMOSTHENES — THE ORATION ON THE CROWN, to which is prefixed **ÆSCHINES AGAINST CTESIPHON.** Edited by B. DRAKE, M.A., late Fellow of King's College, Cambridge. New edition. 5s.

HOMER'S ODYSSEY—THE NARRATIVE OF ODYSSEUS, Books IX.—XII. Edited by JOHN E. B. MAYOR, M.A. Part I. 3s.   [*To be completed shortly.*

JUVENAL—SELECT SATIRES. Edited by JOHN E. B. MAYOR, Fellow of St. John's College, Cambridge, and Professor of Latin. Satires XII.—XVI. 3s. 6d.
                    [*Satires X. and XI., in preparation.*

LIVY—HANNIBAL'S FIRST CAMPAIGN IN ITALY, Books XXI. and XXII. Edited by the Rev. W. W. CAPES, Reader in Ancient History at Oxford. With 3 Maps. 5s.

SALLUST—CATILINE and **JUGURTHA.** Edited by C. MERIVALE, B.D. New edition, carefully revised and enlarged. 4s. 6d. Or separately 2s. 6d. each.

TACITUS—AGRICOLA and **GERMANIA.** Edited by A. J. CHURCH, M.A. and W. J. BRODRIBB, M.A. Translators of Tacitus. New edition. 3s. 6d. Or separately 2s. each.

THE ANNALS, Book VI. By the same Editors. 2s. 6d.

TERENCE—HAUTON TIMORUMENOS. Edited by E. S. SHUCKBURGH, M.A., Assistant-Master at Eton College. 3s. With Translation, 4s. 6d.

THUCYDIDES — THE SICILIAN EXPEDITION, Books VI. and VII. Edited by the Rev. PERCIVAL FROST, M.A., Late Fellow of St. John's College, Cambridge. New edition, revised and enlarged, with Map. 5s.

XENOPHON—HELLENICA, Books I. and II. Edited by H. HAILSTONE, B.A., late Scholar of Peterhouse, Cambridge. With Map. 4s. 6d.

The following are in preparation :—

ÆSCHYLUS—SELECT PLAYS. Edited by A. O. PRICKARD, M.A., Fellow and Tutor of New College, Oxford.

    I. PERSAE.

CATULLUS—SELECT POEMS. Edited by F. P. SIMPSON, B.A., late Scholar of Balliol College, Oxford.

CICERO—PRO ROSCIO AMERINO. From the German of KARL HALM. Edited by E. H. DONKIN, M.A., late Scholar of Lincoln College, Oxford, Assistant Master at Uppingham.

DEMOSTHENES—FIRST PHILIPPIC. Edited by Rev. T. GWATKIN, M.A., late Fellow of St. John's College, Cambridge.

EURIPIDES—SELECT PLAYS, by various Editors.

    ALCESTIS. Edited by J. E. C. WELLDON, B.A., Fellow and Lecturer of King's College, Cambridge.

    BACCHAE. Edited by E. S. SHUCKBURGH, M.A., Assistant-Master at Eton College.

    IPHIGENEIA AT AULIS. Edited by Rev. J. P. MAHAFFY, M.A., Fellow and Tutor of Trinity College, Dublin.

    MEDEA. Edited by A. W. VERRALL, M.A., Fellow and Lecturer of Trinity College, Cambridge.

HERODOTUS—THE INVASION OF GREECE BY XERXES. ooks VII. and VIII. Edited by THOMAS CASE, M.A., formerly Fellow of Brasenose College, Oxford.

HOMER'S ILIAD—THE STORY OF ACHILLES. Edited by the late J. H. PRATT, M.A., and WALTER LEAF, M.A., Fellows of Trinity College, Cambridge.

LYSIAS—SELECT ORATIONS. Edited by E. S. SHUCK-BURGH, M.A., Assistant-Master at Eton College.

MARTIAL—SELECT EPIGRAMS. Edited by Rev. H. M. STEPHENSON, M.A., Head-Master of St. Peter's School, York.

OVID—SELECT EPISTLES. Edited by E. S. SHUCKBURGH, M.A.

OVID—FASTI. Edited by G. H. HALLAM, M.A., Fellow of St. John's College, Cambridge, and Assistant Master at Harrow.

PLATO—FOUR DIALOGUES ON THE TRIAL AND DEATH of SOCRATES, *viz.,* **EUTHYPHRO, APO-LOGY, CRITO, AND PHÆDO.** Edited by C. W. MOULE, M.A., Fellow and Tutor of Corpus Christi College, Cambridge.

PROPERTIUS—SELECT POEMS. Edited by J. P. POST-GATE, B.A., Fellow of Trinity College, Cambridge.

THUCYDIDES—Books I. and II. Edited by H. BROADBENT, M.A., Fellow of Exeter College, Oxford, and Assistant-Master at Eton College.

THUCYDIDES—Books IV. and V. Edited by Rev. C. E. GRAVES, M.A., Classical Lecturer, and late Fellow of St. John's College, Cambridge.

Other volumes will follow.

CLASSICAL.

ÆSCHYLUS—*THE EUMENIDES.* The Greek Text, with Introduction, English Notes, and Verse Translation. By BERNARD DRAKE, M.A., late Fellow of King's College, Cambridge. 8vo. 3s. 6d.

ARISTOTLE—*AN INTRODUCTION TO ARISTOTLE'S RHETORIC.* With Analysis, Notes and Appendices. By E. M. COPE, Fellow and Tutor of Trinity College, Cambridge, 8vo. 14*s.*

ARISTOTLE ON FALLACIES; OR, THE SOPHISTICI ELENCHI. With Translation and Notes by E. POSTE, M.A. Fellow of Oriel College, Oxford. 8vo. 8*s. 6d.*

ARISTOPHANES—*THE BIRDS.* Translated into English Verse, with Introduction, Notes, and Appendices, by B. H. KENNEDY, D.D., Regius Professor of Greek in the University of Cambridge. Crown 8vo. 6*s.* Help-Notes to the same, for the use of Students. 1*s. 6d.*

BELCHER—*SHORT EXERCISES IN LATIN PROSE COMPOSITION AND EXAMINATION PAPERS IN LATIN GRAMMAR*, to which is prefixed a Chapter on Analysis of Sentences. By the Rev. H. BELCHER, M.A., Assistant Master in King's College School, London. New Edition. 18mo. 1*s. 6d.* Key, 1*s. 6d.*

SEQUEL TO THE ABOVE. EXERCISES IN LATIN IDIOMS, &c. By the same author. [*In preparation.*

BLACKIE—*GREEK AND ENGLISH DIALOGUES FOR USE IN SCHOOLS AND COLLEGES.* By JOHN STUART BLACKIE, Professor of Greek in the University of Edinburgh. New Edition. Fcap. 8vo. 2*s. 6d.*

CICERO—*THE ACADEMICA.* The Text revised and explained by JAMES REID, M.A., Fellow of Caius College, Cambridge. New Edition. With Translation. 8vo. [*In preparation.*

SELECT LETTERS.—After the Edition of ALBERT WATSON, M.A. Translated by G. E. JEANS, M.A., Fellow of Hertford College, Oxford, and Assistant-Master at Hailey-bury. [*Shortly.*

CLASSICAL WRITERS. Edited by J. R. GREEN, M.A. Fcap. 8vo. 1s. 6d.

CICERO. By Professor A. S. WILKINS. [*In preparation.*

DEMOSTHENES. By S. H. BUTCHER, M.A. [*In preparation.*

EURIPIDES. By Professor J. P. MAHAFFY. [*In the Press.*

HORACE. By T. H. WARD, M.A. [*In preparation.*

LIVY. By Rev. W. W. CAPES, M.A. [*In preparation.*

VERGIL. By Professor H. NETTLESHIP. [*In preparation.*

Others to follow.

ELLIS—*PRACTICAL HINTS ON THE QUANTITATIVE PRONUNCIATION OF LATIN,* for the use of Classical Teachers and Linguists. By A. J. ELLIS, B.A., F.R.S. Extra fcap. 8vo. 4s. 6d.

GEDDES—*THE PROBLEM OF THE HOMERIC POEMS.* By W. D. GEDDES, Professor of Greek in the University of Aberdeen. 8vo. 14s.

GLADSTONE—Works by the Rt. Hon. W. E. GLADSTONE, M.P. *JUVENTUS MUNDI;* or, Gods and Men of the Heroic Age. Second Edition. Crown 8vo. 10s. 6d.

THE TIME AND PLACE OF HOMER. Crown 8vo. 6s. 6d.

A PRIMER OF HOMER. 18mo. 1s.

GOODWIN—Works by W. W. GOODWIN, Professor of Greek in Harvard University, U.S.A.

SYNTAX OF THE MOODS AND TENSES OF THE GREEK VERB. New Edition, revised. Crown 8vo. 6s. 6d.

AN ELEMENTARY GREEK GRAMMAR. New Edition, revised. Crown 8vo. [*In preparation.*

GREENWOOD—*THE ELEMENTS OF GREEK GRAM-MAR*, including Accidence, Irregular Verbs, and Principles of Derivation and Composition; adapted to the System of Crude Forms. By J. G. GREENWOOD, Principal of Owens College, Manchester. New Edition. Crown 8vo. 5s. 6d.

HERODOTUS, Books I.—III.—*THE EMPIRES OF THE EAST.* Edited, with Notes and Introductions, by A. H. SAYCE, M.A., Fellow and Tutor of Queen's College, Oxford, and Deputy-Professor of Comparative Philology. 8vo.
[In preparation.

HODGSON —*MYTHOLOGY FOR LATIN VERSIFICA-TION.* A brief Sketch of the Fables of the Ancients, prepared to be rendered into Latin Verse for Schools. By F. HODGSON, B.D., late Provost of Eton. New Edition, revised by F. C. HODGSON, M.A. 18mo. 3s.

HOMER—*THE ODYSSEY.* Done into English by S. H. BUTCHER, M.A., Fellow of University College, Oxford, and ANDREW LANG, M.A., late Fellow of Merton College, Oxford, Crown 8vo. 10s. 6d.

HOMERIC DICTIONARY. For Use in Schools and Colleges. Translated from the German of Dr. G. Autenreith, with Additions and Corrections by R. P. KEEP, Ph.D. With numerous Illustrations. Crown 8vo. 6s.

HORACE—*THE WORKS OF HORACE*, rendered into English Prose, with Introductions, Running Analysis, and Notes, by J. LONSDALE, M.A., and S. LEE, M.A. Globe 8vo. 3s. 6d.

THE ODES OF HORACE IN A METRICAL PARA-PHRASE. By R. M. HOVENDEN. Extra fcap. 8vo. 4s.

HORACE'S LIFE AND CHARACTER. An Epitome of his Satires and Epistles. By R. M. HOVENDEN. Extra fcap. 8vo. 4s. 6d.

WORD FOR WORD FROM HORACE. The Odes lite-rally Versified. By W. T. THORNTON, C.B. Crown 8vo. 7s. 6d.

JACKSON—*FIRST STEPS TO GREEK PROSE COM-POSITION.* By BLOMFIELD JACKSON, M.A. Assistant-Master in King's College School, London. New Edition revised and enlarged. 18mo. 1*s*. 6*d*.

JACKSON—*A MANUAL OF GREEK PHILOSOPHY.* By HENRY JACKSON, M.A., Fellow and Prælector in Ancient Philosophy, Trinity College, Cambridge.   [*In preparation.*

JEBB—Works by R. C. JEBB, M.A., Professor of Greek in the University of Glasgow.

THE ATTIC ORATORS FROM ANTIPHON ;TO ISAEOS. 2 vols. 8vo. 25*s*.

THE CHARACTERS OF THEOPHRASTUS. Translated from a revised Text, with Introduction and Notes. Extra fcap. 8vo. 6*s*. 6*d*.

A PRIMER OF GREEK LITERATURE. 18mo. 1*s*.

A HISTORY OF GREEK LITERATURE. Crown 8vo.
[*In preparation.*

JUVENAL—*THIRTEEN SATIRES OF JUVENAL.* With a Commentary. By JOHN E. B. MAYOR, M.A., Kennedy Professor of Latin at Cambridge. Vol. I. Second Edition, enlarged. Crown 8vo. 7*s*. 6*d*. Vol. II. Crown 8vo. 10*s*. 6*d*.

KYNASTON—*GREEK IAMBICS FOR SCHOOLS.* By Rev. H. KYNASTON, M.A., Principal of Cheltenham College.
[*In preparation.*

LIVY, Books XXI.—XXV. Translated by A. J. CHURCH, M.A., and W. J. BRODRIBB, M.A.   [*In preparation.*

LLOYD—*THE AGE OF PERICLES.* A History of the Politics and Arts of Greece from the Persian to the Peloponnesian War. By WILLIAM WATKISS LLOYD. 2 vols. 8vo. 21*s*.

MACMILLAN—*FIRST LATIN GRAMMAR.* By M. C. MACMILLAN, M.A., late Scholar of Christ's College, Cambridge, Assistant Master in St. Paul's School. 18mo. [*In preparation.*

MAHAFFY—Works by J. P. MAHAFFY, M.A., Professor of Ancient History in Trinity College, Dublin.

SOCIAL LIFE IN GREECE ; from Homer to Menander. Third Edition, revised and enlarged. Crown 8vo. 9*s*

RAMBLES AND STUDIES IN GREECE. With Illustrations. Second Edition. With Map. Crown 8vo. 10*s*. 6*d*.

A PRIMER OF GREEK ANTIQUITIES. With Illustrations. 18mo. 1*s*.

MARSHALL — *A TABLE OF IRREGULAR GREEK VERBS*, classified according to the arrangement of Curtius' Greek Grammar. By J. M. MARSHALL, M.A., one of the Masters in Clifton College. 8vo. cloth. New Edition. 1*s*.

MAYOR (JOHN E. B.)—*FIRST GREEK READER.* Edited after KARL HALM, with Corrections and large Additions by Professor JOHN E. B. MAYOR, M.A., Fellow and Classical Lecturer of St. John's College, Cambridge. New Edition, revised. Fcap. 8vo. 4*s*. 6*d*.

BIBLIOGRAPHICAL CLUE TO LATIN LITERATURE. Edited after HÜBNER, with large Additions by Professor JOHN E. B. MAYOR. Crown 8vo. 6*s*. 6*d*.

MAYOR (JOSEPH B.)—*GREEK FOR BEGINNERS.* By the Rev. J. B. MAYOR, M.A., Professor of Classical Literature in King's College, London. Part I., with Vocabulary, 1*s*. 6*d*. Parts II. and III., with Vocabulary and Index, 3*s*. 6*d*. complete in one Vol. New Edition. Fcap. 8vo. cloth. 4*s*. 6*d*.

NIXON—*PARALLEL EXTRACTS* arranged for translation into English and Latin, with Notes on Idioms. By J. E. NIXON, M.A., Classical Lecturer, King's College, London. Part I.—Historical and Epistolary. New Edition, revised and enlarged. Crown 8vo. 3*s*. 6*d*.

NIXON *Continued—*

A FEW NOTES ON LATIN RHETORIC. With Tables and Illustrations. By J. E. NIXON, M.A. Crown 8vo. 2s.

PEILE (JOHN, M.A.)—*AN INTRODUCTION TO GREEK AND LATIN ETYMOLOGY.* By JOHN PEILE, M.A., Fellow and Tutor of Christ's College, Cambridge, formerly Teacher of Sanskrit in the University of Cambridge. Third and Revised Edition. Crown 8vo. 10s. 6d.

A PRIMER OF PHILOLOGY. 18mo. 1s. By the same Author.

PINDAR—*THE EXTANT ODES OF PINDAR.* Translated into English, with an Introduction and short Notes, by ERNEST MYERS, M.A., Fellow of Wadham College, Oxford. Crown 8vo. 5s.

PLATO—*THE REPUBLIC OF PLATO.* Translated into English, with an Analysis and Notes, by J. LL. DAVIES, M.A., and D. J. VAUGHAN, M.A. New Edition, with Vignette Portraits of Plato and Socrates, engraved by JEENS from an Antique Gem. 18mo. 4s. 6d.

PHILEBUS. Edited, with Introduction and Notes, by HENRY JACKSON, M.A., Fellow of Trinity College, Cambridge. 8vo.   [*In preparation.*

PLAUTUS—*THE MOSTELLARIA OF PLAUTUS.* With Notes, Prolegomena, and Excursus. By WILLIAM RAMSAY, M.A., formerly Professor of Humanity in the University of Glasgow. Edited by Professor GEORGE G. RAMSAY, M.A., of the University of Glasgow. 8vo. 14s.

POTTS (A. W., M.A.)—*HINTS TOWARDS LATIN PROSE COMPOSITION.* By ALEXANDER W. POTTS, M.A., LL.D., late Fellow of St. John's College, Cambridge ; Head Master of the Fettes College, Edinburgh. New Edition. Extra fcap. 8vo. 3s.

ROBY—*A GRAMMAR OF THE LATIN LANGUAGE*, from Plautus to Suetonius. By H. J. ROBY, M.A., late Fellow of St. John's College, Cambridge. In Two Parts. Third Edition. Part I. containing :—Book I. Sounds. Book II. Inflexions. Book III. Word-formation. Appendices. Crown 8vo. 8*s*. 6*d*. Part II.—Syntax, Prepositions, &c. Crown 8vo. 10*s*. 6*d*.

"Marked by the clear and practised insight of a master in his art. A book that would do honour to any country."—ATHENÆUM.

SCHOOL LATIN GRAMMAR. By the same Author.
[*In preparation*.

RUSH—*SYNTHETIC LATIN DELECTUS.* A First Latin Construing Book arranged on the Principles of Grammatical Analysis. With Notes and Vocabulary. By E. RUSH, B.A. With Preface by the Rev. W. F. MOULTON, M.A., D.D. Extra fcap. 8vo. [*Immediately*.

RUST—*FIRST STEPS TO LATIN PROSE COMPOSITION.* By the Rev. G. RUST, M.A. of Pembroke College, Oxford, Master of the Lower School, King's College, London. New Edition. 18mo. 1*s*. 6*d*.

RUTHERFORD—*A FIRST GREEK GRAMMAR.* By W. G. RUTHERFORD, M.A., Assistant Master in St. Paul's School, London. Extra fcap. 8vo. 1*s*.

SEELEY—*A PRIMER OF LATIN LITERATURE.* By Prof. J. R. SEELEY. [*In preparation*.

SHUCKBURGH—*A LATIN READER.* By E. S. SHUCK-BURGH, M.A., Assistant Master at Eton College.
[*In preparation*.

TACITUS—*COMPLETE WORKS TRANSLATED.* By A. J. CHURCH, M.A., and W. J. BRODRIBB, M.A.

THE HISTORY. With Notes and a Map. New Edition. Crown 8vo. 6*s*.

THE ANNALS. With Notes and Maps. New Edition. Crown 8vo. 7*s*. 6*d*.

THE AGRICOLA AND GERMANY, WITH THE DIALOGUE ON ORATORY. With Maps and Notes. New and Revised Edition. Crown 8vo. 4*s*. 6*d*.

THEOPHRASTUS—*THE CHARACTERS OF THEO-PHRASTUS.* An English Translation from a Revised Text. With Introduction and Notes. By R. C. JEBB, M.A., Professor of Greek in the University of Glasgow. Extra fcap. 8vo. 6s. 6d.

THRING—Works by the Rev. E. THRING, M.A., Head Master of Uppingham School.

A LATIN GRADUAL. A First Latin Construing Book for Beginners. New Edition, enlarged, with Coloured Sentence Maps. Fcap. 8vo. 2s. 6d.

A MANUAL OF MOOD CONSTRUCTIONS. Fcap. 8vo. 1s. 6d.

A CONSTRUING BOOK. Fcap 8vo. 2s. 6d.

VIRGIL—*THE WORKS OF VIRGIL RENDERED INTO ENGLISH PROSE*, with Notes, Introductions, Running Analysis, and an Index, by JAMES LONSDALE, M.A., and SAMUEL LEE, M.A. New Edition. Globe 8vo. 3s. 6d. ; gilt edges, 4s. 6d.

WILKINS—*A PRIMER OF ROMAN ANTIQUITIES.* By A. S. WILKINS, M.A., Professor of Latin in the Owens College, Manchester. With Illustrations. 18mo. 1s.

WRIGHT—Works by J. WRIGHT, M.A., late Head Master of Sutton Coldfield School.

HELLENICA ; OR, A HISTORY OF GREECE IN GREEK, as related by Diodorus and Thucydides ; being a First Greek Reading Book, with explanatory Notes, Critical and Historical. New Edition with a Vocabulary. Fcap. 8vo. 3s. 6d.

A HELP TO LATIN GRAMMAR; or, The Form and Use of Words in Latin, with Progressive Exercises. Crown 8vo. 4s. 6d.

THE SEVEN KINGS OF ROME. An Easy Narrative, abridged from the First Book of Livy by the omission of Difficult Passages ; being a First Latin Reading Book, with Grammatical Notes. New Edition. With Vocabulary, 3s. 6d.

WRIGHT *Continued—*

FIRST LATIN STEPS; OR, AN INTRODUCTION BY A SERIES OF EXAMPLES TO THE STUDY OF THE LATIN LANGUAGE. Crown 8vo. 5s.

ATTIC PRIMER. Arranged for the Use of Beginners. Extra fcap. 8vo. 4s. 6d.

A COMPLETE LATIN COURSE, comprising Rules with Examples, Exercises, both Latin and English, on each Rule, and Vocabularies. Crown 8vo. 4s. 6d.

MATHEMATICS.

AIRY—Works by Sir G. B. AIRY, K.C.B., Astronomer Royal :—

ELEMENTARY TREATISE ON PARTIAL DIF- FERENTIAL EQUATIONS. Designed for the Use of Students in the Universities. With Diagrams. Second Edition. Crown 8vo. 5s. 6d.

ON THE ALGEBRAICAL AND NUMERICAL THEORY OF ERRORS OF OBSERVATIONS AND THE COMBINATION OF OBSERVATIONS. Second Edition, revised. Crown 8vo. 6s. 6d.

UNDULATORY THEORY OF OPTICS. Designed for the Use of Students in the University. New Edition. Crown 8vo. 6s. 6d.

ON SOUND AND ATMOSPHERIC VIBRATIONS. With the Mathematical Elements of Music. Designed for the Use of Students in the University. Second Edition, Revised and Enlarged. Crown 8vo. 9s.

A TREATISE OF MAGNETISM. Designed for the Use of Students in the University. Crown 8vo. 9s. 6d.

AIRY (OSMUND)—*A TREATISE ON GEOMETRICAL OPTICS.* Adapted for the use of the Higher Classes in Schools. By OSMUND AIRY, B.A., one of the Mathematical Masters in Wellington College. Extra fcap. 8vo. 3s. 6d.

BAYMA—*THE ELEMENTS OF MOLECULAR MECHA-NICS.* By JOSEPH BAYMA, S.J., Professor of Philosophy, Stonyhurst College. Demy 8vo. 10s. 6d.

BEASLEY—*AN ELEMENTARY TREATISE ON PLANE TRIGONOMETRY.* With Examples. By R. D. BEASLEY, M.A., Head Master of Grantham Grammar School. Fifth Edition, revised and enlarged. Crown 8vo. 3s. 6d.

BLACKBURN (HUGH) — *ELEMENTS OF PLANE TRIGONOMETRY,* for the use of the Junior Class in Mathematics n the University of Glasgow. By HUGH BLACKBURN, M.A., Professor of Mathematics in the University of Glasgow. Globe 8vo. 1s. 6d.

BOOLE—Works by G. BOOLE, D.C.L., F.R.S., late Professor of Mathematics in the Queen's University, Ireland.

A TREATISE ON DIFFERENTIAL EQUATIONS. Third and Revised Edition. Edited by I. TODHUNTER. Crown 8vo. 14s.

A TREATISE ON DIFFERENTIAL EQUATIONS. Supplementary Volume. Edited by I. TODHUNTER. Crown 8vo. 8s. 6d.

THE CALCULUS OF FINITE DIFFERENCES. Crown 8vo. 10s. 6d. New Edition, revised by J. F. MOULTON.

BROOK-SMITH (J.)—*ARITHMETIC IN THEORY AND PRACTICE.* By J. BROOK-SMITH, M.A., LL.B., St. John's College, Cambridge; Barrister-at-Law; one of the Masters of Cheltenham College. New Edition, revised. Crown 8vo. 4s. 6d.

CAMBRIDGE SENATE-HOUSE PROBLEMS and RIDERS WITH SOLUTIONS :—

1875—*PROBLEMS AND RIDERS.* By A. G. GREENHILL, M.A. Crown 8vo. 8s. 6d.

1878—*SOLUTIONS OF SENATE-HOUSE PROBLEMS.* By the Mathematical Moderators and Examiners. Edited by J. W. L. GLAISHER, M.A., Fellow of Trinity College, Cambridge. [*In the press.*

CANDLER—*HELP TO ARITHMETIC.* Designed for the use of Schools. By H. CANDLER, M.A., Mathematical Master of Uppingham School. Extra fcap. 8vo. *2s. 6d.*

CHEYNE—*AN ELEMENTARY TREATISE ON THE PLANETARY THEORY.* By C. H. H. CHEYNE, M.A., F.R.A.S. With a Collection of Problems. Second Edition. Crown 8vo. *6s. 6d.*

CHRISTIE—*A COLLECTION OF ELEMENTARY TEST-QUESTIONS IN PURE AND MIXED MATHE-MATICS;* with Answers and Appendices on Synthetic Division, and on the Solution of Numerical Equations by Horner's Method. By JAMES R. CHRISTIE, F.R.S., Royal Military Academy, Woolwich. Crown 8vo. *8s. 6d.*

CLIFFORD—*THE ELEMENTS OF DYNAMIC.* An Introduction to the Study of Motion and Rest in Solid and Fluid Bodies. By A. K. CLIFFORD, F.R.S., Professor of Applied Mathematics and Mechanics at University College, London. Part I.—KINETIC. Crown 8vo. *7s. 6d.*

CUMMING—*AN INTRODUCTION TO THE THEORY OF ELECTRICITY.* By LINNÆUS CUMMING, M.A., one of the Masters of Rugby School. With Illustrations. Crown 8vo. *8s. 6d.*

CUTHBERTSON—*EUCLIDIAN GEOMETRY.* By FRANCIS CUTHBERTSON, M.A., LL.D., Head Mathematical Master of the City of London School. Extra fcap. 8vo. *4s. 6d.*

DALTON—Works by the Rev. T. DALTON, M.A., Assistant Master of Eton College.

RULES AND EXAMPLES IN ARITHMETIC. New Edition. 18mo. *2s. 6d.*

Answers to the Examples are appended.

RULES AND EXAMPLES IN ALGEBRA. Part I. New Edition. 18mo. *2s.* Part II. 18mo. *2s. 6d.*

DAY—*PROPERTIES OF CONIC SECTIONS PROVED GEOMETRICALLY.* PART I., THE ELLIPSE, with Problems. By the Rev. H. G. DAY, M.A. Crown 8vo. 3s. 6d.

DODGSON—*EUCLID AND HIS MODERN RIVALS.* By the Rev. C. L. DODGSON, M.A., Mathematical Lecturer, Christ Church, Oxford. Crown 8vo. [*Nearly ready.*

DREW—*GEOMETRICAL TREATISE ON CONIC SECTIONS.* By W. H. DREW, M.A., St. John's College, Cambridge. New Edition, enlarged. Crown 8vo. 5s.

SOLUTIONS TO THE PROBLEMS IN DREW'S CONIC SECTIONS. Crown 8vo. 4s. 6d.

EDGAR (J. H.) and PRITCHARD (G. S.)—*NOTE-BOOK ON PRACTICAL SOLID OR DESCRIPTIVE GEOMETRY.* Containing Problems with help for Solutions. By J. H. EDGAR, M.A., Lecturer on Mechanical Drawing at the Royal School of Mines, and G. S. PRITCHARD. New Edition, revised and enlarged. Globe 8vo. 3s.

FERRERS—Works by the Rev. N. M. FERRERS, M.A., Fellow and Tutor of Gonville and Caius College, Cambridge.

AN ELEMENTARY TREATISE ON TRILINEAR CO-ORDINATES, the Method of Reciprocal Polars, and the Theory of Projectors. New Edition, revised. Crown 8vo. 6s. 6d.

AN ELEMENTARY TREATISE ON SPHERICAL HARMONICS, AND SUBJECTS CONNECTED WITH THEM. Crown 8vo. 7s. 6d.

FROST—Works by PERCIVAL FROST, M.A., formerly Fellow of St. John's College, Cambridge ; Mathematical Lecturer of King's College.

AN ELEMENTARY TREATISE ON CURVE TRACING. By PERCIVAL FROST, M.A. 8vo. 12s.

SOLID GEOMETRY. A New Edition, revised and enlarged of the Treatise by FROST and WOLSTENHOLME. In 2 Vols. Vol. I. 8vo. 16s.

GODFRAY—Works by HUGH GODFRAY, M.A., Mathematical Lecturer at Pembroke College, Cambridge.

A TREATISE ON ASTRONOMY, for the Use of Colleges and Schools. New Edition. 8vo. 12s. 6d.

AN ELEMENTARY TREATISE ON THE LUNAR THEORY, with a Brief Sketch of the Problem up to the time of Newton. Second Edition, revised. Crown 8vo. 5s. 6d.

HEMMING—*AN ELEMENTARY TREATISE ON THE DIFFERENTIAL AND INTEGRAL CALCULUS,* for the Use of Colleges and Schools. By G. W. HEMMING, M.A., Fellow of St. John's College, Cambridge. Second Edition, with Corrections and Additions. 8vo. 9s.

JACKSON — *GEOMETRICAL CONIC SECTIONS.* An Elementary Treatise in which the Conic Sections are defined as the Plane Sections of a Cone, and treated by the Method of Projection. By J. STUART JACKSON, M.A., late Fellow of Gonville and Caius College, Cambridge. Crown 8vo. 4s. 6d.

JELLET (JOHN H.)—*A TREATISE ON THE THEORY OF FRICTION.* By JOHN H. JELLET, B.D., Senior Fellow of Trinity College, Dublin; President of the Royal Irish Academy. 8vo. 8s. 6d.

JONES and CHEYNE—*ALGEBRAICAL EXERCISES.* Progressively Arranged. By the Rev. C. A. JONES, M.A., and C. H. CHEYNE, M.A., F.R.A.S., Mathematical Masters of Westminster School. New Edition. 18mo. 2s. 6d.

KELLAND and TAIT—*INTRODUCTION TO QUATER-NIONS,* with numerous examples. By P. KELLAND, M.A., F.R.S.; and P. G. TAIT, M.A., Professors in the department of Mathematics in the University of Edinburgh. Crown 8vo. 7s. 6d.

KITCHENER—*A GEOMETRICAL NOTE-BOOK,* containing Easy Problems in Geometrical Drawing preparatory to the Study of Geometry. For the use of Schools. By F. E. KITCHENER, M.A., Mathemathical Master at Rugby. New Edition. 4to. 2s.

b

MAULT—*NATURAL GEOMETRY:* an Introduction to the Logical Study of Mathematics. For Schools and Technical Classes. With Explanatory Models, based upon the Tachymetrical Works of Ed. Lagout. By A. MAULT. 18mo. 1*s.*

Models to Illustrate the above, in Box, 12*s.* 6*d.*

MERRIMAN — *ELEMENTS OF THE METHOD OF LEAST SQUARES.* By MANSFIELD MERRIMAN, Ph.D. Professor of Civic and Mechanical Engineering, Lehigh University, Bethlehem, Penn. Crown 8vo. 7*s.* 6*d.*

MILLAR—*ELEMENTS OF DESCRIPTIVE GEOMETRY.* By J. B. MILLAR, C.E., Assistant Lecturer in Engineering in Owens College, Manchester. Crown 8vo. 6*s.*

MORGAN — *A COLLECTION OF PROBLEMS AND EXAMPLES IN MATHEMATICS.* With Answers. By H. A. Morgan, M.A., Sadlerian and Mathematical Lecturer of Jesus College, Cambridge. Crown 8vo. 6*s.* 6*d.*

MUIR—*DETERMINANTS.* By THOS. MUIR. Crown 8vo. [*In Preparation.*

NEWTON'S *PRINCIPIA.* Edited by Prof. Sir W. THOMSON and Professor BLACKBURN. 4to. cloth. 31*s.* 6*d.*

THE FIRST THREE SECTIONS OF NEWTON'S PRINCIPIA, With Notes and Illustrations. Also a collection of Problems, principally intended as Examples of Newton's Methods. By PERCIVAL FROST, M.A. Third Edition. 8vo. 12*s.*

PARKINSON—Works by S. PARKINSON, D.D., F.R.S., Tutor and Prælector of St. John's College, Cambridge.

AN ELEMENTARY TREATISE ON MECHANICS. For the Use of the Junior Classes at the University and the Higher Classes in Schools. With a Collection of Examples. New Edition, revised. Crown 8vo. cloth. 9*s.* 6*d.*

A TREATISE ON OPTICS. New Edition, revised and enlarged. Crown 8vo. cloth. 10*s.* 6*d.*

PEDLEY—*EXERCISES IN ARITHMETIC.* By S. PEDLEY. [*In preparation.*

PHEAR—*ELEMENTARY HYDROSTATICS.* With Numerous Examples. By J. B. PHEAR, M.A., Fellow and late Assistant Tutor of Clare College, Cambridge. New Edition. Crown 8vo. cloth. 5s. 6d.

PIRIE—*LESSONS ON RIGID DYNAMICS.* By the Rev. G. PIRIE, M.A., Fellow and Tutor of Queen's College, Cambridge. Crown 8vo. 6s.

PUCKLE—*AN ELEMENTARY TREATISE ON CONIC SECTIONS AND ALGEBRAIC GEOMETRY.* With Numerous Examples and Hints for their Solution; especially designed for the Use of Beginners. By G. H. PUCKLE, M.A. New Edition, revised and enlarged. Crown 8vo. 7s. 6d.

RAWLINSON—*ELEMENTARY STATICS*, by the Rev. GEORGE RAWLINSON, M.A. Edited by the Rev. EDWARD STURGES, M.A. Crown 8vo. 4s. 6d.

RAYLEIGH—*THE THEORY OF SOUND.* By LORD RAYLEIGH, M.A., F.R.S., formerly Fellow of Trinity College, Cambridge. 8vo. Vol. I. 12s. 6d. Vol. II. 12s. 6d.
[*Vol. III. in the Press.*

REYNOLDS—*MODERN METHODS IN ELEMENTARY GEOMETRY.* By E. M. REYNOLDS, M.A., Mathematical Master in Clifton College. Crown 8vo. 3s. 6d.

ROUTH—Works by EDWARD JOHN ROUTH, M.A., F.R.S., late Fellow and Assistant Tutor of St. Peter's College, Cambridge; Examiner in the University of London.

' *AN ELEMENTARY TREATISE ON THE DYNAMICS OF THE SYSTEM OF RIGID BODIES.* With numerous Examples. Third and enlarged Edition. 8vo. 21s.

STABILITY OF A GIVEN STATE OF MOTION, PARTICULARLY STEADY MOTION. Adams' Prize Essay for 1877. 8vo. 8s. 6d.

b 2

SMITH—Works by the Rev. BARNARD SMITH, M.A., Rector of Glaston, Rutland, late Fellow and Senior Bursar of St. Peter's College, Cambridge.

ARITHMETIC AND ALGEBRA, in their Principles and Application ; with numerous systematically arranged Examples taken from the Cambridge Examination Papers, with especial reference to the Ordinary Examination for the B.A. Degree. New Edition, carefully revised. Crown 8vo. 10s. 6d.

ARITHMETIC FOR SCHOOLS. New Edition. Crown 8vo. 4s. 6d.

A KEY TO THE ARITHMETIC FOR SCHOOLS. New Edition. Crown 8vo. 8s. 6d.

EXERCISES IN ARITHMETIC. Crown 8vo. limp cloth. 2s. With Answers. 2s. 6d.

Or sold separately, Part I. 1s. ; Part II. 1s. ; Answers, 6d.

SCHOOL CLASS-BOOK OF ARITHMETIC. 18mo. cloth. 3s.

Or sold separately, in Three Parts. 1s. each.

KEYS TO SCHOOL CLASS-BOOK OF ARITHMETIC. Parts I., II., and III., 2s. 6d. each.

SHILLING BOOK OF ARITHMETIC FOR NATIONAL AND ELEMENTARY SCHOOLS. 18mo. cloth. Or separately, Part I. 2d. ; Part II. 3d. ; Part III. 7d. Answers. 6d.

THE SAME, with Answers complete. 18mo, cloth. 1s. 6d.

KEY TO SHILLING BOOK OF ARITHMETIC. 18mo. 4s. 6d.

EXAMINATION PAPERS IN ARITHMETIC. 18mo. 1s. 6d. The same, with Answers, 18mo. 2s. Answers, 6d.

KEY TO EXAMINATION PAPERS IN ARITH-METIC. 18mo. 4s. 6d.

SMITH *Continued*—

THE METRIC SYSTEM OF ARITHMETIC, ITS PRINCIPLES AND APPLICATIONS, with numerous Examples, written expressly for Standard V. in National Schools. New Edition. 18mo. cloth, sewed. *3d*.

A CHART OF THE METRIC SYSTEM, on a Sheet, size 42 in. by 34 in. on Roller, mounted and varnished price *3s. 6d.* New Edition.

Also a Small Chart on a Card, price *1d*.

EASY LESSONS IN ARITHMETIC, combining Exercises in Reading, Writing, Spelling, and Dictation. Part I. for Standard I. in National Schools. Crown 8vo. *9d*.

EXAMINATION CARDS IN ARITHMETIC. (Dedicated to Lord Sandon.) With Answers and Hints.

Standards I. and II. in box, *1s*. Standards III., IV. and V., in boxes, *1s*. each. Standard VI. in Two Parts, in boxes, *1s*. each.

A and B papers, of nearly the same difficulty, are given so as to prevent copying, and the Colours of the A and B papers differ in each Standard, and from those of every other Standard, so that a master or mistress can see at a glance whether the children have the proper papers.

SNOWBALL —. *THE ELEMENTS OF PLANE AND SPHERICAL TRIGONOMETRY;* with the Construction and Use of Tables of Logarithms. By J. C. SNOWBALL, M.A. New Edition. Crown 8vo. *7s. 6d*.

SYLLABUS OF PLANE GEOMETRY (corresponding to Euclid, Books I.—VI.). Prepared by the Association for the Improvement of Geometrical Teaching. New Edition. Crown 8vo. *1s*.

TAIT and STEELE—*A TREATISE ON DYNAMICS OF A PARTICLE.* With numerous Examples. By Professor TAIT and MR. STEELE. Fourth Edition, revised. Crown 8vo. *12s*.

TEBAY—*ELEMENTARY MENSURATION FOR SCHOOLS.* With numerous Examples. By SEPTIMUS TEBAY, B.A., Head Master of Queen Elizabeth's Grammar School, Rivington. Extra fcap. 8vo. 3*s.* 6*d.*

TODHUNTER—Works by I. TODHUNTER, M.A., F.R.S., of St. John's College, Cambridge.

> ' Mr. Todhunter is chiefly known to students of Mathematics as the author of a series of admirable mathematical text-books, which possess the rare qualities of being clear in style and absolutely free from mistakes, typographical or other."—SATURDAY REVIEW.

THE ELEMENTS OF EUCLID. For the Use of Colleges and Schools. New Edition. 18mo. 3*s.* 6*d.*

MENSURATION FOR BEGINNERS. With numerous Examples. New Edition. 18mo. 2*s.* 6*d.*

ALGEBRA FOR BEGINNERS. With numerous Examples. New Edition. 18mo. 2*s.* 6*d.*

KEY TO ALGEBRA FOR BEGINNERS. Crown 8vo. 6*s.* 6*a.*

TRIGONOMETRY FOR BEGINNERS. With numerous Examples. New Edition. 18mo. 2*s.* 6*d.*

KEY TO TRIGONOMETRY FOR BEGINNERS. Crown 8vo. 8*s.* 6*d.*

MECHANICS FOR BEGINNERS. With numerous Examples. New Edition. 18mo. 4*s.* 6*d.*

KEY TO MECHANICS FOR BEGINNERS. Crown 8vo. 6*s.* 6*d.*

ALGEBRA. For the Use of Colleges and Schools. New Edition. Crown 8vo. 7*s.* 6*d.*

KEY TO ALGEBRA FOR THE USE OF COLLEGES AND SCHOOLS. Crown 8vo. 10*s.* 6*d.*

AN ELEMENTARY TREATISE ON THE THEORY OF EQUATIONS. New Edition, revised. Crown 8vo. 7*s.* 6*d.*

TODHUNTER *Continued—*

PLANE TRIGONOMETRY. For Schools and Colleges. New Edition. Crown 8vo. 5*s.*

KEY TO PLANE TRIGONOMETRY. Crown 8vo. 10*s.* 6*d.*

A TREATISE ON SPHERICAL TRIGONOMETRY. New Edition, enlarged. Crown 8vo. 4*s.* 6*d.*

PLANE CO-ORDINATE GEOMETRY, as applied to the Straight Line and the Conic Sections. With numerous Examples. New Edition, revised and enlarged. Crown 8vo. 7*s.* 6*d.*

A TREATISE ON THE DIFFERENTIAL CALCULUS. With numerous Examples. New Edition. Crown 8vo. 10*s.* 6*d.*

A TREATISE ON THE INTEGRAL CALCULUS AND ITS APPLICATIONS. With numerous Examples. New Edition, revised and enlarged. Crown 8vo. 10*s.* 6*d.*

EXAMPLES OF ANALYTICAL GEOMETRY OF THREE DIMENSIONS. New Edition, revised. Crown 8vo. 4*s.*

A TREATISE ON ANALYTICAL STATICS. With numerous Examples. New Edition, revised and enlarged. Crown 8vo. 10*s.* 6*d.*

A HISTORY OF THE MATHEMATICAL THEORY OF PROBABILITY, from the time of Pascal to that of Laplace. 8vo. 18*s.*

RESEARCHES IN THE CALCULUS OF VARIA-TIONS, principally on the Theory of Discontinuous Solutions : an Essay to which the Adams Prize was awarded in the University of Cambridge in 1871. 8vo. 6*s.*

TODHUNTER *Continued—*

A HISTORY OF THE MATHEMATICAL THEORIES OF ATTRACTION, AND THE FIGURE OF THE EARTH, from the time of Newton to that of Laplace. 2 vols. 8vo. 24s.

AN ELEMENTARY TREATISE ON LAPLACE'S, LAME'S, AND BESSEL'S FUNCTIONS. Crown 8vo. 10s. 6d.

WILSON (J. M.)—*ELEMENTARY GEOMETRY.* Books I. to V. Containing the Subjects of Euclid's first Six Books. Following the Syllabus of the Geometrical Association. By J. M. WILSON, M.A., Head Master of Clifton College. New Edition. Extra fcap. 8vo. 4s. 6d.

SOLID GEOMETRY AND CONIC SECTIONS. With Appendices on Transversals and Harmonic Division. For the Use of Schools. By J. M. WILSON, M.A. New Edition. Extra fcap. 8vo. 3s. 6d.

WILSON—*GRADUATED EXERCISES IN PLANE TRI-GONOMETRY.* Compiled and arranged by J. WILSON, M.A., and S. R. WILSON, B.A. Crown 8vo. [*Immediately.*

WILSON (W. P.)—*A TREATISE ON DYNAMICS.* By W. P. WILSON, M.A., Fellow of St. John's College, Cambridge, and Professor of Mathematics in Queen's College, Belfast. 8vo. 9s. 6d.

WOLSTENHOLME—*MATHEMATICAL PROBLEMS,* on Subjects included in the First and Second Divisions of the Schedule of Subjects for the Cambridge Mathematical Tripos Examination. Devised and arranged by JOSEPH WOLSTEN-HOLME, late Fellow of Christ's College, sometime Fellow of St. John's College, and Professor of Mathematics in the Royal Indian Engineering College. New Edition greatly enlarged. 8vo. 18s.

SCIENCE.

SCIENCE PRIMERS FOR ELEMENTARY SCHOOLS.

Under the joint Editorship of Professors HUXLEY, ROSCOE, and BALFOUR STEWART.

> "These Primers are extremely simple and attractive, and thoroughly answer their purpose of just leading the young beginner up to the threshold of the long avenues in the Palace of Nature which these titles suggest."
> —GUARDIAN.
> "They are wonderfully clear and lucid in their instruction, simple in style, and admirable in plan.'—EDUCATIONAL TIMES.

CHEMISTRY — By H. E. ROSCOE, F.R.S., Professor of Chemistry in Owens College, Manchester. With numerous Illustrations. 18mo. 1s. New Edition. With Questions.

> "A very model of perspicacity and accuracy."—CHEMIST AND DRUG-GIST.

PHYSICS—By BALFOUR STEWART, F.R.S., Professor of Natural Philosophy in Owens College, Manchester. With numerous Illustrations. 18mo. 1s. New Edition. With Questions.

PHYSICAL GEOGRAPHY—By ARCHIBALD GEIKIE, F.R.S. Murchison Professor of Geology and Mineralogy at Edinburgh. With numerous Illustrations. New Edition, with Questions. 18mo. 1s.

> "Everyone of his lessons is marked by simplicity, clearness, and correctness."—ATHENÆUM.

GEOLOGY — By Professor GEIKIE, F.R.S. With numerous Illustrations. New Edition. 18mo. cloth. 1s.

> "It is hardly possible for the dullest child to misunderstand the meaning of a classification of stones after Professor Geikie's explanation."—SCHOOL BOARD CHRONICLE.

PHYSIOLOGY—By MICHAEL FOSTER, M.D., F.R.S. With numerous Illustrations. New Edition. 18mo. 1s.

> "The book seems to us to leave nothing to be desired as an elementary text-book."—ACADEMY.

SCIENCE PRIMERS *Continued*—

ASTRONOMY —.By J. NORMAN LOCKYER, F.R.S. With numerous Illustrations. New Edition. 18mo. 1*s*.

> "This is altogether one of the most likely attempts we have ever seen to bring astronomy down to the capacity of the young child."—SCHOOL BOARD CHRONICLE.

BOTANY—By Sir J. D. HOOKER, K.C.S.I., C B., President of the Royal Society. With numerous Illustrations. New Edition. 18mo. 1*s*.

> "To teachers the Primer will be of inestimable value, and not only because of the simplicity of the language and the clearness with which the subject matter is treated, but also on account of its coming from the highest authority, and so furnishing positive information as to the most suitable methods of teaching the science of botany."—NATURE.

LOGIC—By Professor STANLEY JEVONS, F.R.S. New Edition. 18mo. 1*s*.

> "It appears to us admirably adapted to serve both as an introduction to scientific reasoning, and as a guide to sound judgment and reasoning in the ordinary affairs of life."—ACADEMY.

POLITICAL ECONOMY—By Professor STANLEY JEVONS, F.R.S. 18mo. 1*s*.

> "Unquestionably in every respect an admirable primer."—SCHOOL BOARD CHRONICLE.

In preparation :—

INTRODUCTORY. By Professor HUXLEY. &c. &c.

ELEMENTARY CLASS-BOOKS.

ASTRONOMY, by the Astronomer Royal.
POPULAR ASTRONOMY. With Illustrations. By Sir G. B. AIRY, K.C.B., Astronomer Royal. New Edition. 18mo. 4*s*. 6*d*.

ASTRONOMY.
ELEMENTARY LESSONS IN ASTRONOMY. With Coloured Diagram of the Spectra of the Sun, Stars, and Nebulæ, and numerous Illustrations. By J. NORMAN LOCKYER, F.R.S. New Edition. Fcap. 8vo. 5*s*. 6*d*.

> "Full, clear, sound, and worthy of attention, not only as a popular exposition, but as a scientific ' Index.' "—ATHENÆUM.

ELEMENTARY CLASS-BOOKS *Continued—*

QUESTIONS ON LOCKYER'S ELEMENTARY LESSONS IN ASTRONOMY. For the Use of Schools. By JOHN FORBES-ROBERTSON. 18mo. cloth limp. 1s. 6d.

PHYSIOLOGY.

LESSONS IN ELEMENTARY PHYSIOLOGY. With numerous Illustrations. By T. H. HUXLEY, F.R.S., Professor of Natural History in the Royal School of Mines. New Edition. Fcap. 8vo. 4s. 6d.

"Pure gold throughout."—GUARDIAN.
"Unquestionably the clearest and most complete elementary treatise on this subject that we possess in any language."—WESTMINSTER REVIEW.

QUESTIONS ON HUXLEY'S PHYSIOLOGY FOR SCHOOLS. By T. ALCOCK, M.D. 18mo. 1s. 6d.

BOTANY.

LESSONS IN ELEMENTARY BOTANY. By D. OLIVER, F.R.S., F.L.S., Professor of Botany in University College, London. With nearly Two Hundred Illustrations New Edition. Fcap. 8vo. 4s. 6d.

CHEMISTRY.

LESSONS IN ELEMENTARY CHEMISTRY, INORGANIC AND ORGANIC. By HENRY E. ROSCOE, F.R.S., Professor of Chemistry in Owens College, Manchester. With numerous Illustrations and Chromo-Litho of the Solar Spectrum, and of the Alkalies and Alkaline Earths. New Edition. Fcap. 8vo. 4s. 6d.

"As a standard general text-book it deserves to take a leading place."—SPECTATOR.
"We unhesitatingly pronounce it the best of all our elementary treatises on Chemistry."—MEDICAL TIMES.

A SERIES OF CHEMICAL PROBLEMS, prepared with Special Reference to the above, by T. E. Thorpe, Ph.D., Professor of Chemistry in the Yorkshire College of Science, Leeds. Adapted for the preparation of Students for the Government, Science, and Society of Arts Examinations. With a Preface by Professor ROSCOE. Fifth Edition, with Key, 18mo. 2s.

ELEMENTARY CLASS-BOOKS *Continued—*

POLITICAL ECONOMY.

POLITICAL ECONOMY FOR BEGINNERS. By MILLICENT G. FAWCETT. New Edition. 18mo. 2s. 6d.

" Clear, compact, and comprehensive."—DAILY NEWS.
" The relations of capital and labour have never been more simply or more clearly expounded."—CONTEMPORARY REVIEW.

LOGIC.

ELEMENTARY LESSONS IN LOGIC; Deductive and Inductive, with copious Questions and Examples, and a Vocabulary of Logical Terms. By W. STANLEY JEVONS, M.A., Professor of Political Economy in University College, London. New Edition. Fcap. 8vo. 3s. 6d.

" Nothing can be better for a school-book."—GUARDIAN.
" A manual alike simple, interesting, and scientific."—ATHENÆUM.

PHYSICS.

LESSONS IN ELEMENTARY PHYSICS. By BALFOUR STEWART, F.R.S., Professor of Natural Philosophy in Owens College, Manchester. With numerous Illustrations and Chromolitho of the Spectra of the Sun, Stars, and Nebulæ. New Edition. Fcap. 8vo. 4s. 6d.

" The beau-ideal of a scientific text-book, clear, accurate, and thorough."
EDUCATIONAL TIMES.

PRACTICAL CHEMISTRY.

THE OWENS COLLEGE JUNIOR COURSE OF PRACTICAL CHEMISTRY. By FRANCIS JONES, Chemical Master in the Grammar School, Manchester. With Preface by Professor ROSCOE, and Illustrations. New Edition. 18mo. 2s. 6d.

ANATOMY.

LESSONS IN ELEMENTARY ANATOMY. By ST. GEORGE MIVART, F.R.S., Lecturer in Comparative Anatomy at St. Mary's Hospital. With upwards of 400 Illustrations. Fcap. 8vo. 6s. 6d.

" It may be questioned whether any other work on anatomy contains in like compass so proportionately great a mass of information."—LANCET.
" The work is excellent, and should be in the hands of every student of human anatomy."—MEDICAL TIMES.

ELEMENTARY CLASS-BOOKS *Continued—*
MECHANICS.

AN ELEMENTARY TREATISE. By A. B. W.
KENNEDY, C.E., Professor of Applied Mechanics in University
College, London. With Illustrations. . [*In preparation.*

STEAM.

AN ELEMENTARY TREATISE. By JOHN PERRY,
Professor of Engineering, Imperial College of Engineering,
Yedo. With numerous Woodcuts and Numerical Examples
and Exercises. 18mo. 4s. 6d.

"The young engineer and those seeking for a comprehensive knowledge
of the use, power, and economy of steam, could not have a more useful
work, as it is very intelligible, well arranged, and practical throughout."—
IRONMONGER.

PHYSICAL GEOGRAPHY.

*ELEMENTARY LESSONS IN PHYSICAL GEO-
GRAPHY.* By A. GEIKIE, F.R.S., Murchison Professor
of Geology, &c., Edinburgh. With numerous Illustrations.
Fcap. 8vo. 4s. 6d.

QUESTIONS ON THE SAME. 1s. 6d.

GEOGRAPHY.

CLASS-BOOK OF GEOGRAPHY. By C. B. Clarke, M.A.,
F.R.G.S. Fcap. 8vo. 2s. 6d.

NATURAL PHILOSOPHY.

NATURAL PHILOSOPHY FOR BEGINNERS. By
I. TODHUNTER, M.A., F.R.S. Part I. The Properties of
Solid and Fluid Bodies. 18mo. 3s. 6d.
Part II. Sound, Light, and Heat. 18mo. 3s. 6d.

SOUND—*AN ELEMENTARY TREATISE.* By W.H. STONE,
M.D., F.R.S. With Illustrations. 18mo. [*In the Press.*

Others in Preparation.

MANUALS FOR STUDENTS.

Crown 8vo.

DYER AND VINES—*THE STRUCTURE OF PLANTS.* By Professor THISELTON DYER, F.R.S., assisted by SYDNEY VINES, B.Sc., Fellow and Lecturer of Christ's College, Cambridge.   With numerous Illustrations.   [*In preparation.*

FAWCETT—*A MANUAL OF POLITICAL ECONOMY.* By Professor FAWCETT, M.P.   New Edition, revised and enlarged.   Crown 8vo.   12*s.* 6*d.*

FLEISCHER—*A SYSTEM OF VOLUMETRIC ANALY-SIS.*   Translated, with Notes and Additions, from the second German Edition, by M. M. PATTISON MUIR, F.R.S.E.   With Illustrations.   Crown 8vo.   7*s.* 6*d.*

FLOWER (W. H.)—*AN INTRODUCTION TO THE OSTE-OLOGY OF THE MAMMALIA.*   Being the substance of the Course of Lectures delivered at the Royal College of Surgeons of England in 1870.   By Professor W. H. FLOWER, F.R.S., F.R.C.S.   With numerous Illustrations.   New Edition, enlarged.   Crown 8vo.   10*s.* 6*d.*

FOSTER and BALFOUR—*THE ELEMENTS OF EMBRYO-LOGY.*   By MICHAEL FOSTER, M.D., F.R.S., and F. M. BALFOUR, M.A.   Part I. crown 8vo.   7*s.* 6*d.*

FOSTER and LANGLEY—*A COURSE OF ELEMENTARY PRACTICAL PHYSIOLOGY.*   By MICHAEL FOSTER, M.D., F.R.S., and J. N. LANGLEY, B.A.   New Edition. Crown 8vo.   6*s.*

HOOKER (Dr.)—*THE STUDENTS FLORA OF THE BRITISH ISLANDS.*   By Sir J. D. HOOKER, K.C.S.I., C.B., F.R.S., M.D., D.C.L.   New Edition, revised.   Globe 8vo,   10*s.* 6*d.*

MANUALS FOR STUDENTS *Continued*—

HUXLEY—*PHYSIOGRAPHY.* An Introduction to the Study of Nature. By Professor HUXLEY, F.R.S. With numerous Illustrations, and Coloured Plates. New Edition. Crown 8vo. 7s.6d.

HUXLEY and MARTIN—*A COURSE OF PRACTICAL INSTRUCTION IN ELEMENTARY BIOLOGY.* By Professor HUXLEY, F.R.S., assisted by H. N. MARTIN, M.B., D.Sc. New Edition, revised. Crown 8vo. 6s.

HUXLEY and PARKER—*ELEMENTARY BIOLOGY. PART II.* By Professor HUXLEY, F.R.S., assisted by — PARKER. With Illustrations. [*In preparation.*

JEVONS—*THE PRINCIPLES OF SCIENCE.* A Treatise on Logic and Scientific Method. By Professor W. STANLEY JEVONS, LL.D., F.R.S. New and Revised Edition. Crown 8vo. 12s. 6d.

OLIVER (Professor)—*FIRST BOOK OF INDIAN BOTANY.* By Professor DANIEL OLIVER, F.R.S., F.L.S., Keeper of the Herbarium and Library of the Royal Gardens, Kew, With numerous Illustrations. Extra fcap. 8vo. 6s. 6d.

PARKER and BETTANY—*THE MORPHOLOGY OF THE SKULL.* By Professor PARKER and G. T. BETTANY. Illustrated. Crown 8vo. 10s. 6d.

TAIT—*AN ELEMENTARY TREATISE ON HEAT.* By Professor TAIT, F.R.S.E. Illustrated. [*In the Press.*

THOMSON—*ZOOLOGY.* By Sir C. WYVILLE THOMSON, F.R.S. Illustrated. [*In preparation.*

TYLOR and LANKESTER—*ANTHROPOLOGY.* By E. B. TYLOR, M.A., F.R.S., and Professor E. RAY LANKESTER, M.A., F.R.S. Illustrated. [*In preparation.*

Other volumes of these Manuals will follow.

SCIENTIFIC TEXT-BOOKS.

BALL (R. S., A.M.)—*EXPERIMENTAL MECHANICS.* A Course of Lectures delivered at the Royal College of Science for Ireland. By R. S. BALL, A.M., Professor of Applied Mathematics and Mechanics in the Royal College of Science for Ireland. Royal 8vo. 16*s.*

FOSTER—*A TEXT BOOK OF PHYSIOLOGY.* By MICHAEL FOSTER, M.D., F.R.S. With Illustrations. New Edition, enlarged, with additional Illustrations. 8vo. 21*s.*

GAMGEE —*A TEXT-BOOK, SYSTEMATIC AND PRACTICAL, OF THE PHYSIOLOGICAL CHEMISTRY OF THE ANIMAL BODY.* Including the changes which the Tissues and Fluids undergo in Disease. By A. GAMGEE, M.D., F.R.S., Professor of Physiology, Owens College, Manchester. 8vo.   [*In preparation.*

GEGENBAUR—*ELEMENTS OF COMPARATIVE ANATOMY.* By Professor CARL GEGENBAUR. A Translation by F. JEFFREY BELL, B.A. Revised with Preface by Professor E. RAY LANKESTER, F.R.S. With numerous Illustrations. 8vo. 21*s.*

KLAUSIUS—*MECHANICAL THEORY OF HEAT.* Translated by WALTER K. BROWNE. 8vo.   [*In preparation.*

NEWCOMB—*POPULAR ASTRONOMY.* By S. NEWCOMB, LL.D., Professor U.S. Naval Observatory. With 112 Illustrations and 5 Maps of the Stars. 8vo. 18*s.*
  "It is unlike anything else of its kind, and will be of more use in circulating a knowledge of astronomy than nine-tenths of the books which have appeared on the subject of late years."—*Saturday Review*.

REULEAUX — *THE KINEMATICS OF MACHINERY.* Outlines of a Theory of Machines. By Professor F. REULEAUX. Translated and Edited by Professor A. B. KENNEDY, C.E. With 450 Illustrations. Medium 8vo. 21*s.*

SCIENTIFIC TEXT-BOOKS *Continued*—

ROSCOE and SCHORLEMMER—*CHEMISTRY*, A Complete Treatise on. By Professor H. E. ROSCOE, F.R.S., and Professor C. SCHORLEMMER, F.R.S. Medium 8vo. Vol. I.— The Non-Metallic Elements. With numerous Illustrations, and Portrait of Dalton. 21s. Vol. II.—Metals. Part I. Illustrated. 18s. [*Vol. II.—Metals. Part II. in the Prsss.*

SCHORLEMMER—*A MANUAL OF THE CHEMISTRY OF THE CARBON COMPOUNDS, OR ORGANIC CHEMISTRY.* By C. SCHORLEMMER, F.R.S., Professor of Chemistry, Owens College, Manchester. With Illustrations. 8vo. 14s.

NATURE SERIES.

THE SPECTROSCOPE AND ITS APPLICATIONS. By J. NORMAN LOCKYER, F.R.S. With Coloured Plate and numerous Illustrations. Second Edition. Crown 8vo. 3s. 6d.

THE ORIGIN AND METAMORPHOSES OF INSECTS. By Sir JOHN LUBBOCK, M.P., F.R.S., D.C.L. With numerous Illustrations. Second Edition. Crown 8vo. 3s. 6d.

THE TRANSIT OF VENUS. By G. FORBES, M.A., Professor of Natural Philosophy in the Andersonian University, Glasgow. Illustrated. Crown 8vo. 3s. 6d.

THE COMMON FROG. By ST. GEORGE MIVART, F.R.S., Lecturer in Comparative Anatomy at St. Mary's Hospital. With numerous Illustrations. Crown 8vo. 3s. 6d.

POLARISATION OF LIGHT. By W. SPOTTISWOODE, F.R.S. With many Illustrations. Second Edition. Crown 8vo. 3s. 6d.

ON BRITISH WILD FLOWERS CONSIDERED IN RELATION TO INSECTS. By Sir JOHN LUBBOCK, M.P., F.R.S. With numerous Illustrations. Second Edition. Crown 8vo. 4s. 6d.

THE SCIENCE OF WEIGHING AND MEASURING, AND THE STANDARDS OF MEASURE AND WEIGHT. By H. W. CHISHOLM, Warden of the Standards. With numerous Illustrations. Crown 8vo. 4s. 6d.

c

NATURE SERIES *Continued—*

HOW TO DRAW A STRAIGHT LINE : a Lecture on Link-
ages. By A. B. KEMPE. With Illustrations. Crown 8vo. 1s. 6d.

LIGHT: a Series of Simple, entertaining, and Inexpensive Experiments in the Phenomena of Light, for the Use of Students of
every age. By A. M. MAYER and C. BARNARD. Crown 8vo,
with numerous Illustrations. 2s. 6d.

SOUND : a Series of Simple, Entertaining, and Inexpensive Experiments in the Phenomena of Sound, for the use of Students
of every age. By A. M. MAYER, Professor of Physics in
the Stevens Institute of Technology, &c. With numerous
Illustrations. Crown 8vo. 3s. 6d.

FIELD GEOLOGY. By Prof. GEIKIE, F.R.S., Director of the
Geological Survey of Scotland.   [*In the Press.*

Other volumes to follow.

EASY LESSONS IN SCIENCE.

HEAT. By Miss C. A. MARTINEAU. Edited by Prof. W. F.
BARRETT.   [*In preparation.*

LIGHT. By Mrs. AWDRY. Edited by Prof. W. F. BARRETT.
                                          [*In preparation.*

ELECTRICITY. By Prof. W. F. BARRETT. [*In preparation.*

SCIENCE LECTURES AT SOUTH KENSINGTON.

VOL. I. Containing Lectures by Capt. ABNEY, Prof. STOKES,
Prof. KENNEDY, F. G. BRAMWELL, Prof. G. FORBES, H. C.
SORBY, J. T. BOTTOMLEY, S. H. VINES, and Prof. CAREY
FOSTER. Crown 8vo. 6s.

VOL. II. Containing Lectures by W. SPOTTISWOODE, P.R.S.,
Prof. FORBES, Prof. PIGOT, Prof. BARRETT, Dr. BURDON-
SANDERSON, Dr. LAUDER BRUNTON, F.R.S., Prof. ROSCOE,
and others. Crown 8vo. 6s.

MANCHESTER SCIENCE LECTURES FOR THE PEOPLE.

Eighth Series, 1876-7. Crown 8vo. Illustrated. 6*d*. each.

WHAT THE EARTH IS COMPOSED OF. By Professor ROSCOE, F.R.S.

THE SUCCESSION OF LIFE ON THE EARTH. By Professor WILLIAMSON, F.R.S.

WHY THE EARTH'S CHEMISTRY IS AS IT IS. By J. N. LOCKYER, F.R.S.

Also complete in One Volume. Crown 8vo. cloth. 2*s*.

BLANFORD—*THE RUDIMENTS OF PHYSICAL GEO-GRAPHY FOR THE USE OF INDIAN SCHOOLS;* with a Glossary of Technical Terms employed. By H. F. BLANFORD, F.R.S. New Edition, with Illustrations. Globe 8vo. 2*s*. 6*d*

GORDON—*AN ELEMENTARY BOOK ON HEAT.* By J. E. H. GORDON, B.A., Gonville and Caius College, Cambridge. Crown 8vo. 2*s*.

M'KENDRICK—*OUTLINES OF PHYSIOLOGY IN ITS RELATIONS TO MAN.* By J. G. M'KENDRICK, M.D., F.R.S.E. With Illustrations. Crown 8vo. 12*s*. 6*d*.

MIALL—*STUDIES IN COMPARATIVE ANATOMY.*
No. I.—The Skull of the Crocodile : a Manual for Students. By L. C. MIALL, Professor of Biology in the Yorkshire College and Curator of the Leeds Museum. 8vo. 2*s*. 6*d*.

No. II.—Anatomy of the Indian Elephant. By L. C. MIALL and F. GREENWOOD. With Illustrations. 8vo. 5*s*.

MUIR—*PRACTICAL CHEMISTRY FOR MEDICAL STU-DENTS.* Specially arranged for the first M.B. Course. By M. M. PATTISON MUIR, F.R.S.E. Fcap. 8vo. 1*s*. 6*d*.

SHANN—*AN ELEMENTARY TREATISE ON HEAT, IN RELATION TO STEAM AND THE STEAM-ENGINE.* By G. SHANN, M.A. With Illustrations. Crown 8vo. 4*s*. 6*d*.

WRIGHT—*METALS AND THEIR CHIEF INDUSTRIAL APPLICATIONS.* By C. ALDER WRIGHT, D.Sc., &c. Lecturer on Chemistry in St. Mary's Hospital Medica School. Extra fcap. 8vo. 3*s*. 6*d*.

c

HISTORY.

BEESLY—*STORIES FROM THE HISTORY OF ROME.* By Mrs. BEESLY. Fcap. 8vo. 2s. 6d.

"The attempt appears to us in every way successful. The stories are interesting in themselves, and are told with perfect simplicity and good feeling."—DAILY NEWS.

FREEMAN (EDWARD A.)—*OLD-ENGLISH HISTORY.* By EDWARD A. FREEMAN, D.C.L., LL.D., late Fellow of Trinity College, Oxford. With Five Coloured Maps. New Edition. Extra fcap. 8vo. half-bound. 6s.

GREEN—*A SHORT HISTORY OF THE ENGLISH PEOPLE.* By JOHN RICHARD GREEN. With Coloured Maps, Genealogical Tables, and Chronological Annals. Crown 8vo. 8s. 6d. Fifty-fifth Thousand.

"Stands alone as the one general history of the country, for the sake of which all others, if young and old are wise, will be speedily and surely set aside."—ACADEMY.

HISTORICAL COURSE FOR SCHOOLS—Edited by EDWARD A. FREEMAN, D.C.L., late Fellow of Trinity College, Oxford.

I. *GENERAL SKETCH OF EUROPEAN HISTORY.* By EDWARD A. FREEMAN, D.C.L. New Edition, revised and enlarged, with Chronological Table, Maps, and Index. 18mo. cloth. 3s. 6d.

"It supplies the great want of a good foundation for historical teaching. The scheme is an excellent one, and this instalment has been executed in a way that promises much for the volumes that are yet to appear."—EDUCATIONAL TIMES.

II. *HISTORY OF ENGLAND.* By EDITH THOMPSON. New Edition, revised and enlarged, with Maps. 18mo. 2s. 6d.

III. *HISTORY OF SCOTLAND.* By MARGARET MACARTHUR. New Edition. 18mo. 2s.

"An excellent summary, unimpeachable as to facts, and putting them in the clearest and most impartial light attainable."—GUARDIAN.

IV. *HISTORY OF ITALY.* By the Rev. W. HUNT, M.A. 18mo. 3s.

"It possesses the same solid merit as its predecessors . . . . the same scrupulous care about fidelity in details. . . . It is distinguished, too, by information on art, architecture, and social politics, in which the writer's grasp is seen by the firmness and clearness of his touch"—EDUCATIONAL TIMES.

HISTORICAL COURSE FOR SCHOOLS *Continued—*

V. *HISTORY OF GERMANY.*   By J. SIMÉ, M.A. 18mo. 3*s.*

> "A remarkably clear and impressive history of Germany. Its great events are wisely kept as central figures, and the smaller events are carefully kept, not only subordinate and subservient, but most skilfully woven into the texture of the historical tapestry presented to the eye."— STANDARD.

VI. *HISTORY OF AMERICA.*   By JOHN A. DOYLE. With Maps. 18mo. 4*s.* 6*d.*

> "Mr. Doyle has performed his task with admirable care, fulness, and clearness, and for the first time we have for schools an accurate and interesting history of America, from the earliest to the present time."— STANDARD.

EUROPEAN COLONIES.   By E. J. PAYNE, M.A.   With Maps. 18mo. 4*s.* 6*d.*

> "We have seldom met with an historian capable of forming a more comprehensive, far-seeing, and unprejudiced estimate of events and peoples, and we can commend this little work as one certain to prove of the highest interest to all thoughtful readers."—TIMES.

FRANCE.   By CHARLOTTE M. YONGE.   [*In preparation.*

GREECE.   By EDWARD A. FREEMAN, D.C.L.

                          [*In preparation.*

ROME.   By EDWARD A. FREEMAN, D.C.L.   [*In preparation.*

HISTORY PRIMERS—Edited by JOHN RICHARD GREEN. Author of "A Short History of the English People."

ROME.   By the Rev. M. CREIGHTON, M.A., Fellow and Tutor of Merton College, Oxford. With Eleven Maps. 18mo. 1*s.*

> "The author has been curiously successful in telling in an intelligent way the story of Rome from first to last."—SCHOOL BOARD CHRONICLE.

GREECE.   By C. A. FYFFE, M.A., Fellow and late Tutor of University College, Oxford. With Five Maps. 18mo. 1*s.*

> "We give our unqualified praise to this little manual."—SCHOOLMASTER.

EUROPEAN HISTORY.   By E. A. FREEMAN, D.C.L., LL.D. With Maps. 18mo. 1*s.*

> "The work is always clear, and forms a luminous key to European history."—SCHOOL BOARD CHRONICLE.

HISTORY PRIMERS *Continued—*

GREEK ANTIQUITIES. By the Rev. J. P. MAHAFFY, M.A. Illustrated. 18mo. 1s.

"All that is necessary for the scholar to know is told so compactly yet so fully, and in a style so interesting, that it is impossible for even the dullest boy to look on this little work in the same light as he regards his other school books."—SCHOOLMASTER.

CLASSICAL GEOGRAPHY. By H. F. TOZER, M.A. 18mo. 1s.

"Another valuable aid to the study of the ancient world. . . . It contains an enormous quantity of information packed into a small space, and at the same time communicated in a very readable shape."—JOHN BULL.

GEOGRAPHY. By GEORGE GROVE, D.C.L. With Maps. 18mo. 1s.

"A model of what such a work should be . . . . we know of no short treatise better suited to infuse life and spirit into the dull lists of proper names of which our ordinary class-books so often almost exclusively consist."—TIMES.

ROMAN ANTIQUITIES. By Professor WILKINS. Illustrated. 18mo. 1s.

"A little book that throws a blaze of light on Roman History, and is, moreover, intensely interesting."—*School Board Chronicle.*

FRANCE. By CHARLOTTE M. YONGE. 18mo. 1s.

In preparation :—

ENGLAND. By J. R. GREEN, M.A.

MICHELET—*A SUMMARY OF MODERN HISTORY.* Translated from the French of M. MICHELET, and continued to the Present Time, by M. C. M. SIMPSON. Globe 8vo. 4s. 6d.

OTTÉ—*SCANDINAVIAN HISTORY.* By E. C. OTTÉ. With Maps. Globe 8vo. 6s.

PAULI—*PICTURES OF OLD ENGLAND.* By Dr. R. PAULI. Translated with the sanction of the Author by E. C. OTTÉ. Cheaper Edition. Crown 8vo. 6s.

TAIT—*ANALYSIS OF ENGLISH HISTORY*, based on Green's "Short History of the English People." By C. W. A. TAIT, M.A., Assistant Master, Clifton College. Crown 8vo. 3s. 6d.

YONGE (CHARLOTTE M.)—*A PARALLEL HISTORY OF FRANCE AND ENGLAND* : consisting of Outlines and Dates. By CHARLOTTE M. YONGE, Author of "The Heir of Redclyffe," &c., &c. Oblong 4to. 3s. 6d.

CAMEOS FROM ENGLISH HISTORY. — FROM ROLLO TO EDWARD II. By the Author of "The Heir of Redclyffe." Extra fcap. 8vo. New Edition. 5s.

A SECOND SERIES OF CAMEOS FROM ENGLISH HISTORY—THE WARS IN FRANCE. New Edition. Extra fcap. 8vo. 5s.

A THIRD SERIES OF CAMEOS FROM ENGLISH HISTORY—THE WARS OF THE ROSES. New Edition. Extra fcap. 8vo. 5s.

A FOURTH SERIES. [*In the press.*

EUROPEAN HISTORY. Narrated in a Series of Historical Selections from the Best Authorities. Edited and arranged by E. M. SEWELL and C. M. YONGE. First Series, 1003—1154. Third Edition. Crown 8vo. 6s. Second Series, 1088—1228. New Edition. Crown 8vo. 6s.

DIVINITY.

. For other Works by these Authors, see THEOLOGICAL CATALOGUE.

ABBOTT (REV. E. A.)—*BIBLE LESSONS.* By the Rev. E. A. ABBOTT, D.D., Head Master of the City of London School. New Edition. Crown 8vo. 4s. 6d.

"Wise, suggestive, and really profound initiation into religious thought."—GUARDIAN.

ARNOLD—*A BIBLE-READING FOR SCHOOLS*—THE GREAT PROPHECY OF ISRAEL'S RESTORATION (Isaiah, Chapters xl.—lxvi.). Arranged and Edited for Young Learners. By MATTHEW ARNOLD, D.C.L., formerly Professor of Poetry in the University of Oxford, and Fellow of Oriel. New Edition. 18mo. cloth. 1s.

ARNOLD *Continued*—:

ISAIAH XL.—LXVI. With the Shorter Prophecies allied to it. Arranged and Edited, with Notes, by MATTHEW ARNOLD. Crown 8vo. 5*s*.

GOLDEN TREASURY PSALTER—Students' Edition. Being an Edition of "The Psalms Chronologically Arranged, by Four Friends," with briefer Notes. 18mo. 3*s*. 6*d*.

GREEK TESTAMENT. Edited, with Introduction and Appendices, by CANON WESTCOTT and Dr. F. J. A. HORT. Two Vols. Crown 8vo. [*In the press.*

HARDWICK—Works by Archdeacon HARDWICK.

A HISTORY OF THE CHRISTIAN CHURCH. Middle Age. From Gregory the Great to the Excommunication of Luther. Edited by WILLIAM STUBBS, M.A., Regius Professor of Modern History in the University of Oxford. With Four Maps constructed for this work by A. KEITH JOHNSTON. Fourth Edition. Crown 8vo. 10*s*. 6*d*.

A HISTORY OF THE CHRISTIAN CHURCH DURING THE REFORMATION. Fourth Edition. Edited by Professor STUBBS. Crown 8vo. 10*s*. 6*d*.

KING—*CHURCH HISTORY OF IRELAND.* By the Rev. ROBERT KING. New Edition. 2 vols. Crown 8vo. [*In preparation.*

MACLEAR—Works by the Rev. G. F. MACLEAR, D.D., Head Master of King's College School.

A CLASS-BOOK OF OLD TESTAMENT HISTORY. New Edition, with Four Maps. 18mo. 4*s*. 6*d*.

A CLASS-BOOK OF NEW TESTAMENT HISTORY, including the Connection of the Old and New Testament. With Four Maps. New Edition. 18mo. 5*s*. 6*d*.

A SHILLING BOOK OF OLD TESTAMENT HISTORY, for National and Elementary Schools. With Map. 18mo. cloth. New Edition.

A SHILLING BOOK OF NEW TESTAMENT HISTORY, for National and Elementary Schools. With Map. 18mo. cloth. New Edition.

MACLEAR *Continued—*

These works have been carefully abridged from the author's larger manuals.

CLASS-BOOK OF THE CATECHISM OF THE CHURCH OF ENGLAND. New Edition. 18mo. cloth. 1*s.* 6*d.*

A FIRST CLASS-BOOK OF THE CATECHISM OF THE CHURCH OF ENGLAND, with Scripture Proofs, for Junior Classes and Schools. 18mo. 6*d.* New Edition.

A MANUAL OF INSTRUCTION FOR CONFIRMATION AND FIRST COMMUNION. WITH PRAYERS AND. DEVOTIONS. 32mo. cloth extra, red edges. 2*s.*

M'CLELLAN—*THE NEW TESTAMENT.* A New Translation on the Basis of the Authorised Version, from a Critically revised Greek Text, with Analyses, copious References and Illustrations from original authorities, New Chronological and Analytical Harmony of the Four Gospels, Notes and Dissertations. A contribution to Christian'Evidence. By JOHN BROWN M'CLELLAN, M.A., late Fellow of Trinity College, Cambridge. In Two Vols. Vol. I.—The Four Gospels with the Chronological and Analytical Harmony. 8vo. 30*s.*

> "One of the most remarkable productions of recent times," says the *Theological Review*, "in this department of sacred literature;" and the *British Quarterly Review* terms it "a thesaurus of first-hand investigations."

MAURICE—*THE LORD'S PRAYER, THE CREED, AND THE COMMANDMENTS.* Manual for Parents and Schoolmasters. To which is added the Order of the Scriptures. By the Rev. F. DENISON MAURICE, M.A. 18mo. cloth, limp. 1*s.*

PROCTER—*A HISTORY OF THE BOOK OF COMMON PRAYER*, with a Rationale of its Offices. By FRANCIS PROCTER, M.A. Thirteenth Edition, revised and enlarged. Crown 8vo. 10*s.* 6*d.*

PROCTER AND MACLEAR—*AN ELEMENTARY INTRO-DUCTION TO THE BOOK OF COMMON PRAYER.* Re-arranged and supplemented by an Explanation of the Morning and Evening Prayer and the Litany. By the Rev. F. PROCTER and the Rev. Dr. MACLEAR. New and Enlarged Edition, containing the Communion Service and the Confirmation and Baptismal Offices. 18mo. 2*s.* 6*d.*

PSALMS OF DAVID CHRONOLOGICALLY ARRANGED. By Four Friends. An Amended Version, with Historical Introduction and Explanatory Notes. Second and Cheaper Edition, with Additions and Corrections. Cr. 8vo. 8*s.* 6*d.*

RAMSAY—*THE CATECHISER'S MANUAL;* or, the Church Catechism Illustrated and Explained, for the Use of Clergymen, Schoolmasters, and Teachers. By the Rev. ARTHUR RAMSAY, M.A. New Edition. 18mo. 1*s.* 6*d.*

SIMPSON—*AN EPITOME OF THE HISTORY OF THE CHRISTIAN CHURCH.* By WILLIAM SIMPSON, M.A. New Edition. Fcap. 8vo. 3*s.* 6*d.*

TRENCH—By R. C. TRENCH, D.D., Archbishop of Dublin. *LECTURES ON MEDIEVAL CHURCH HISTORY.* Being the substance of Lectures delivered at Queen's College, London. Second Edition, revised. 8vo. 12*s.*

SYNONYMS OF THE NEW TESTAMENT. Eighth Edition, revised. 8vo. 12*s.*

WESTCOTT—Works by BROOKE FOSS WESTCOTT, D.D., Canon of Peterborough.

A GENERAL SURVEY OF THE HISTORY OF THE CANON OF THE NEW TESTAMENT DURING THE FIRST FOUR CENTURIES. Fourth Edition. With Preface on "Supernatural Religion." Crown 8vo. 10*s.* 6*d.*

INTRODUCTION TO THE STUDY OF THE FOUR GOSPELS. Fifth Edition. Crown 8vo. 10*s.* 6*d.*

WESTCOTT *Continued*—

THE BIBLE IN THE CHURCH. A Popular Account of the Collection and Reception of the Holy Scriptures in the Christian Churches. New Edition. 18mo. cloth. 4*s*. 6*d*.

THE GOSPEL OF THE RESURRECTION. Thoughts on its Relation to Reason and History. New Edition. Crown 8vo. 6*s*.

WILSON—*THE BIBLE STUDENT'S GUIDE* to the more Correct Understanding of the English Translation of the Old Testament, by reference to the original Hebrew. By WILLIAM WILSON, D.D., Canon of Winchester, late Fellow of Queen's College, Oxford. Second Edition, carefully revised. 4to. cloth. 25*s*.

YONGE (CHARLOTTE M.)—*SCRIPTURE READINGS FOR SCHOOLS AND FAMILIES.* By CHARLOTTE M. YONGE, Author of "The Heir of Redclyffe."

FIRST SERIES. GENESIS TO DEUTERONOMY. Globe 8vo. 1*s*. 6*d*. With Comments. 3*s*. 6*d*.

SECOND SERIES. From JOSHUA to SOLOMON. Extra fcap. 8vo. 1*s*. 6*d*. With Comments, 3*s*. 6*d*.

THIRD SERIES. The KINGS and the PROPHETS. Extra fcap. 8vo. 1*s*. 6*d*. With Comments, 3*s*. 6*d*.

FOURTH SERIES. The GOSPEL TIMES. 1*s*. 6*d*. With Comments, extra fcap. 8vo., 3*s*. 6*d*.

FIFTH SERIES. [*In the press.*

MISCELLANEOUS.

Including works on English, French, and German Languages and Literature, Art Hand-books, &c., &c.

ABBOTT—*A SHAKESPEARIAN GRAMMAR.* An Attempt to illustrate some of the Differences between Elizabethan and Modern English. By the Rev. E. A. ABBOTT, D.D., Head Master of the City of London School. New Edition. Extra fcap. 8vo. 6*s*.

ANDERSON—*LINEAR PERSPECTIVE, AND MODEL DRAWING.* A School and Art Class Manual, with Questions and Exercises for Examination, and Examples of Examination Papers. By LAURENCE ANDERSON. With Illustrations. Royal 8vo. 2s.

BARKER—*FIRST LESSONS IN THE PRINCIPLES OF COOKING.* By LADY BARKER. New Edition. 18mo. 1s.

BEAUMARCHAIS—*LE BARBIER DE SEVILLE.* Edited, with Introduction and Notes, by L. P. BLOUET, Assistant Master in St. Paul's School. Fcap. 8vo. 3s. 6d.

BERNERS—*FIRST LESSONS ON HEALTH.* By J. BERNERS. New Edition. 18mo. 1s.

BLAKISTON—*THE TEACHER.* Practical Suggestions for the improvement of Primary Instruction. By J. R. BLAKISTON, M.A., H.M. Inspector of Schools. Crown 8vo.
[*Immediately.*

BREYMANN—Works by HERMANN BREYMANN, Ph.D., Professor of Philology in the University of Munich.

A FRENCH GRAMMAR BASED ON PHILOLOGICAL PRINCIPLES. Second Edition. Extra fcap. 8vo. 4s. 6d.

FIRST FRENCH EXERCISE BOOK. Extra fcap. 8vo. 4s. 6d.

SECOND FRENCH EXERCISE BOOK. Extra fcap. 8vo. 2s. 6d.

CALDERWOOD—*HANDBOOK OF MORAL PHILOSOPHY.* By the Rev. HENRY CALDERWOOD, LL.D., Professor of Moral Philosophy, University of Edinburgh. New Edition. Crown 8vo. 6s.

DELAMOTTE—*A BEGINNER'S DRAWING BOOK.* By P. H. DELAMOTTE, F.S.A. Progressively arranged. New Edition improved. Crown 8vo. 3s. 6d.

FAWCETT—*TALES IN POLITICAL ECONOMY.* By MILLICENT GARRETT FAWCETT. Globe 8vo. 3s.

FEARON—*SCHOOL INSPECTION.* By D. R. FEARON, M.A., Assistant Commissioner of Endowed Schools. Third Edition. Crown 8vo. 2*s.* 6*d.*

GLADSTONE—*SPELLING REFORM FROM AN EDU-CATIONAL POINT OF VIEW.* By J. H. GLADSTONE, F.R.S., Member for the School Board for London. New Edition. Crown 8vo. 1*s.* 6*d.*

GOLDSMITH—*THE TRAVELLER,* or a Prospect of Society; and *THE DESERTED VILLAGE.* By OLIVER GOLD-SMITH. With Notes Philological and Explanatory, by J. W. HALES, M.A. Crown 8vo. 6*d.*

GREEN—*READINGS FROM ENGLISH HISTORY.* Selected and Edited by JOHN RICHARD GREEN, M.A., LL.D., Honorary Fellow of Jesus College, Oxford. Three Parts. Globe 8vo. 1*s.* 6*d.* each. [*Shortly.*

HALES—*LONGER ENGLISH POEMS,* with Notes, Philological and Explanatory, and an Introduction on the Teaching of English. Chiefly for Use in Schools. Edited by J. W. HALES, M.A., Professor of English Literature at King's College, London, &c. &c. New Edition. Extra fcap. 8vo. 4*s.* 6*d.*

HOLE—*A GENEALOGICAL STEMMA OF THE KINGS OF ENGLAND AND FRANCE.* By the Rev. C. HOLE. On Sheet. 1*s.*

JOHNSON'S *LIVES OF THE POETS.* The Six Chief Lives (Milton, Dryden, Swift, Addison, Pope, Gray), with Macaulay's "Life of Johnson." Edited with Preface by MATTHEW ARNOLD. Crown 8vo. 6*s.*

LITERATURE PRIMERS—Edited by JOHN RICHARD GREEN, Author of "A Short History of the English People."

ENGLISH GRAMMAR. By the Rev. R. MORRIS, LL.D., sometime President of the Philological Society. 18mo. cloth. 1*s.*

LITERATURE PRIMERS *Continued—*

ENGLISH GRAMMAR EXERCISES. By R. MORRIS, LL.D., and H. C. BOWEN, M.A. 18mo. 1*s.*

THE CHILDREN'S TREASURY OF LYRICAL POETRY. Selected and arranged with Notes by FRANCIS TURNER PALGRAVE. In Two Parts. 18mo. 1*s.* each.

ENGLISH LITERATURE. By the Rev. STOPFORD BROOKE, M.A. New Edition. 18mo. 1*s.*

PHILOLOGY. By J. PEILE, M.A. 18mo. 1*s.*

GREEK LITERATURE. By Professor JEBB, M.A. 18mo. 1*s.*

SHAKSPERE. By Professor DOWDEN. 18mo. 1*s.*

HOMER. By the Right Hon. W. E. GLADSTONE, M.P. 18mo. 1*s.*

ENGLISH COMPOSITION. By Professor NICHOL. 18mo. 1*s.*

In preparation :—

GEOGRAPHY OF GREAT BRITAIN AND IRE-LAND. By J. R. GREEN, and ALICE STOPFORD GREEN.
                                          [*Nearly ready.*

LATIN LITERATURE. By Professor SEELEY.

HISTORY OF THE ENGLISH LANGUAGE. By J. A. H. MURRAY, LL.D.

MACMILLAN'S COPY-BOOKS—

Published in two sizes, viz. :—

    1. Large Post 4to. Price 4*d.* each.

    2. Post Oblong. Price 3*d.* each.

*1. INITIATORY EXERCISES & SHORT LETTERS

*2. WORDS CONSISTING OF SHORT LETTERS.

MACMILLAN'S COPY-BOOKS *Continued—*

*3. *LONG LETTERS.* With words containing Long Letters—Figures.

*4. *WORDS CONTAINING LONG LETTERS.*

4a. *PRACTISING AND REVISING COPY-BOOK.* For Nos. 1 to 4.

*5. *CAPITALS AND SHORT HALF-TEXT.* Words beginning with a Capital.

*6. *HALF-TEXT WORDS,* beginning with a Capital—Figures.

*7. *SMALL-HAND AND HALF-TEXT.* With Capitals and Figures.

*8. *SMALL-HAND AND HALF-TEXT.* With Capitals and Figures.

8a. *PRACTISING AND REVISING COPY-BOOK.* For Nos. 5 to 8.

*9. *SMALL-HAND SINGLE HEADLINES*—Figures.

10. *SMALL-HAND SINGLE HEADLINES*—Figures.

*11. *COMMERCIAL AND ARITHMETICAL EXAMPLES, &c.*

12a. *PRACTISING AND REVISING COPY-BOOK.* For Nos. 8 to 12.

* *These numbers may be had with Goodman's Patent Sliding Copies.* Large Post 4to. Price 6d. each.

By a simple device the copies, which are printed upon separate slips, are arranged with a movable attachment, by which they are adjusted so as to be directly before the eye of the pupil at

MACMILLAN'S COPY-BOOKS *Continued*—

all points of his progress. It enables him, also, to keep his own faults concealed, with perfect models constantly in view for imitation. Every experienced teacher knows the advantage of the slip copy, but its practical application has never before been successfully accomplished. This feature is secured exclusively to Macmillan's Copy-books under Goodman's patent.

An inspection of books written on the old plan, with copies at the head of the page, will show that the lines last written at the bottom are almost invariably the poorest. The copy has been too far from the pupil's eye to be of any practical use, and a repetition and exaggeration of his errors have been the result.

MACMILLAN'S PROGRESSIVE FRENCH COURSE—By G. EUGENE-FASNACHT, Senior Master of Modern Languages, Harpur Foundation Modern School, Bedford.

I.—FIRST YEAR, containing Easy Lessons on the Regular Accidence. Extra fcap. 8vo. 1*s.*

II.—SECOND YEAR, containing Conversational Lessons on Systematic Accidence and Elementary Syntax. With Philological Illustrations and Etymological Vocabulary. 1*s.* 6*d.*

MACMILLAN'S PROGRESSIVE GERMAN COURSE—By G. EUGENE FASNACHT.

Part I.—FIRST YEAR. Easy Lessons and Rules on the Regular Accidence. Extra fcap. 8vo. 1*s.* 6*d.*

Part II.—SECOND YEAR. Conversational Lessons in Systematic Accidence and Elementary Syntax. With Philological Illustrations and Etymological Vocabulary. Extra fcap. 8vo. 2*s.*

MARTIN — *THE POET'S HOUR:* Poetry selected and arranged for Children. By FRANCES MARTIN. Third Edition. 18mo. 2*s.* 6*d.*

SPRING-TIME WITH THE POETS: Poetry selected by FRANCES MARTIN. Second Edition. 18mo. 3*s.* 6*d.*

MASSON (GUSTAVE)—*A COMPENDIOUS DICTIONARY OF THE FRENCH LANGUAGE* (French-English and English-French). Followed by a List of the Principal Diverging Derivations, and preceded by Chronological and Historical Tables. By GUSTAVE MASSON, Assistant-Master and Librarian, Harrow School. Fourth Edition. Crown 8vo. half-bound. 6*s*.

MORRIS—Works by the Rev. R. MORRIS, LL.D., Lecturer on English Language and Literature in King's College School.

HISTORICAL OUTLINES OF ENGLISH ACCIDENCE, comprising Chapters on the History and Development of the Language, and on Word-formation. New Edition. Extra fcap. 8vo. 6*s*.

ELEMENTARY LESSONS IN HISTORICAL ENGLISH GRAMMAR, containing Accidence and Word-formation. New Edition. 18mo. 2*s*. 6*d*.

PRIMER OF ENGLISH GRAMMAR. 18mo. 1*s*.

NICOL—*HISTORY OF THE FRENCH LANGUAGE*, with especial reference to the French element in English. By HENRY NICOL, Member of the Philological Society.

[*In preparation.*

OLIPHANT—*THE OLD AND MIDDLE ENGLISH.* A New Edition of "*THE SOURCES OF STANDARD ENGLISH*," revised and greatly enlarged. By T. KINGTON OLIPHANT. Extra fcap. 8vo. 9*s*.

PALGRAVE—*THE CHILDREN'S TREASURY OF LYRICAL POETRY.* Selected and Arranged with Notes by FRANCIS TURNER PALGRAVE. 18mo. 2*s*. 6*d*. Also in Two parts. 18mo. 1*s*. each.

PLUTARCH—Being a Selection from the Lives which Illustrate Shakespeare. North's Translation. Edited, with Introductions, Notes, Index of Names, and Glossarial Index, by the Rev. W. W. SKEAT, M.A. Crown 8vo. 6*s*.

d

PYLODET—*NEW GUIDE TO GERMAN CONVERSA-TION:* containing an Alphabetical List of nearly 800 Familiar Words followed by Exercises, Vocabulary of Words in frequent use ; Familiar Phrases and Dialogues ; a Sketch of German Literature, Idiomatic Expressions, &c. By L. PYLODET. 18mo. cloth limp. 2s. 6d.

A SYNOPSIS OF GERMAN GRAMMAR. From the above. 18mo. 6d.

READING BOOKS—Adapted to the English and Scotch Codes. Bound in Cloth.

PRIMER. 18mo. (48 pp.) 2d.

BOOK I. for Standard I. 18mo. (96 pp.) 4d.
  „ II. „ II. 18mo. (144 pp.) 5d.
  „ III. „ III. 18mo. (160 pp.) 6d.
  „ IV. „ IV. 18mo. (176 pp.) 8d.
  „ V. „ V. 18mo. (380 pp.) 1s.
  „ VI. „ VI. Crown 8vo. (430 pp.) 2s.

Book VI. is fitted for higher Classes, and as an Introduction to English Literature.

> "They are far above any others that have appeared both in form and substance. . . . The editor of the present series has rightly seen that reading books must 'aim chiefly at giving to the pupils the power of accurate, and, if possible, apt and skilful expression ; at cultivating in them a good literary taste, and at arousing a desire of further reading. This is done by taking care to select the extracts from true English classics, going up in Standard VI. course to Chaucer, Hooker, and Bacon, as well as Wordsworth, Macaulay, and Froude. . . . This is quite on the right track, and indicates justly the ideal which we ought to set before us."— GUARDIAN.

SHAKESPEARE—*A SHAKESPEARE MANUAL.* By F. G. FLEAY, M.A., Head Master of Skipton Grammar School. Second Edition. Extra fcap. 8vo. 4s. 6d.

AN ATTEMPT TO DETERMINE THE CHRONO-LOGICAL ORDER OF SHAKESPEARE'S PLAYS. By the Rev. H. PAINE STOKES, B.A. Extra fcap. 8vo. 4s. 6d.

THE TEMPEST. With Glossarial and Explanatory Notes. By the Rev. J. M. JEPHSON. Second Edition. 18mo. 1s.

SONNENSCHEIN and MEIKLEJOHN — *THE ENGLISH METHOD OF TEACHING TO READ.* By A. SONNENSCHEIN and J. M. D. MEIKLEJOHN, M.A. Fcap. 8vo.

COMPRISING :

THE NURSERY BOOK, containing all the Two-Letter Words in the Language. 1*d.* (Also in Large Type on Sheets for School Walls. 5*s.*)

THE FIRST COURSE, consisting of Short Vowels with Single Consonants. 6*d.*

THE SECOND COURSE, with Combinations and Bridges. consisting of Short Vowels with Double Consonants. 6*d.*

THE THIRD AND FOURTH COURSES, consisting of Long Vowels, and all the Double Vowels in the Language. 6*d.*

"These are admirable books, because they are constructed on a principle, and that the simplest principle on which it is possible to learn to read English."—SPECTATOR.

TANNER—*FIRST PRINCIPLES OF AGRICULTURE.* By H. TANNER, F.C.S., Professor of Agricultural Science, University College, Aberystwith, &c. 18mo. 1*s.*

TAYLOR—*WORDS AND PLACES;* or, Etymological Illustrations of History, Ethnology, and Geography. By the Rev. ISAAC TAYLOR, M.A. Third and cheaper Edition, revised and compressed. With Maps. Globe 8vo. 6*s.*

A HISTORY OF THE ALPHABET. By the same Author. [*In preparation.*

TAYLOR—*A PRIMER OF PIANOFORTE PLAYING.* By FRANKLIN TAYLOR. Edited by GEORGE GROVE. 18mo. 1*s*

TEGETMEIER — *HOUSEHOLD MANAGEMENT AND COOKERY.* With an Appendix of Recipes used by the Teachers of the National School of Cookery. By W. B. TEGETMEIER. Compiled at the request of the School Board for London. 18mo. 1*s.*

THRING—Works by EDWARD THRING, M.A., Head Master of Uppingham.

THE ELEMENTS OF GRAMMAR TAUGHT IN ENGLISH. With Questions. Fourth Edition. 18mo. 2s.

THE CHILD'S GRAMMAR. Being the Substance of "The Elements of Grammar taught in English," adapted for the Use of Junior Classes. A New Edition. 18mo. 1s.

SCHOOL SONGS. A Collection of Songs for Schools. With the Music arranged for four Voices. Edited by the Rev. E. THRING and H. RICCIUS. Folio. 7s. 6d.

TRENCH (ARCHBISHOP)—Works by R. C. TRENCH, D.D., Archbishop of Dublin.

HOUSEHOLD BOOK OF ENGLISH POETRY. Selected and Arranged, with Notes. Second Edition. Extra fcap. 8vo. 5s. 6d.

ON THE STUDY OF WORDS. Lectures addressed (originally) to the Pupils at the Diocesan Training School, Winchester. Seventeenth Edition, revised. Fcap. 8vo. 5s.

ENGLISH, PAST AND PRESENT. Tenth Edition, revised and improved. Fcap. 8vo. 5s.

A SELECT GLOSSARY OF ENGLISH WORDS, used formerly in Senses Different from their Present. Fcap. 8vo. 4s. 6d.   [*New Edition in the Press.*

VAUGHAN (C. M.)—*WORDS FROM THE POETS.* By C. M. VAUGHAN. Eighth Edition. 18mo. cloth. 1s.

WEIR—*HARRISON WEIR'S DRAWING COPY-BOOKS.* Oblong 4to. I. Animals.   [*In preparation.*

WHITNEY—Works by WILLIAM D. WHITNEY, Professor of Sanskrit and Instructor in Modern Languages in Yale College; first President of the American Philological Association, and hon. member of the Royal Asiatic Society of Great Britain and Ireland; and Correspondent of the Berlin Academy of Sciences.

A COMPENDIOUS GERMAN GRAMMAR. Crown 8vo. 4s. 6d.

WHITNEY *Continued—*

A GERMAN READER IN PROSE AND VERSE, with Notes and Vocabulary. Crown 8vo. 5*s.*

WRIGHT—*SCHOOL COOKERY.* Edited by C. E. GUTHRIE WRIGHT, Hon. Sec. to the Edinburgh School of Cookery. 18mo. *[Shortly.*

WHITNEY AND EDGREN—*A COMPENDIOUS GERMAN AND ENGLISH DICTIONARY*, with Notation of Correspondences and Brief Etymologies. By Professor W. D. WHITNEY, assisted by A. H. EDGREN. Crown 8vo. 7*s.* 6*d.*

THE GERMAN-ENGLISH PART, separately, 5*s.*

YONGE (CHARLOTTE M.)—*THE ABRIDGED BOOK OF GOLDEN DEEDS.* A Reading Book for Schools and general readers. By the Author of "The Heir of Redclyffe." 18mo. cloth. 1*s.*

MACMILLAN'S
GLOBE LIBRARY.

Beautifully printed on toned paper, price 3s. 6d. each. Also kept in various morocco and calf bindings, at moderate prices.

The *Saturday Review* says :—"The Globe Editions are admirable for their scholarly editing, their typographical excellence, their compendious form, and their cheapness."

The *Daily Telegraph* calls it "a series yet unrivalled for its combination of excellence and cheapness."

SHAKESPEARE'S COMPLETE WORKS. Edited by W. G. CLARK, M.A., and W. ALDIS WRIGHT, M.A. With Glossary.

MORTE D'ARTHUR. Sir Thomas Malory's Book of King Arthur and of his Noble Knights of the Round Table. The Edition of Caxton, revised for Modern Use. With an Introduction, Notes, and Glossary, by Sir EDWARD STRACHEY.

BURNS'S COMPLETE WORKS: the Poems, Songs, and Letters. Edited, with Glossarial Index and Biographical Memoir, by ALEXANDER SMITH.

ROBINSON CRUSOE. Edited after the Original Editions, with Biographical Introduction, by HENRY KINGSLEY.

SCOTT'S POETICAL WORKS. With Biographical and Critica Essay, by FRANCIS TURNER PALGRAVE.

GOLDSMITH'S MISCELLANEOUS WORKS. With Biographical Introduction by Professor MASSON.

SPENCER'S COMPLETE WORKS. Edited, with Glossary, by R. MORRIS, and Memoir by J. W. HALES.

POPE'S POETICAL WORKS. Edited, with Notes and Introductory Memoir, by Professor WARD.

DRYDEN'S POETICAL WORKS. Edited, with a Revised Text and Notes, by W. D. CHRISTIE, M.A., Trinity College, Cambridge.

COWPER'S POETICAL WORKS. Edited, with Notes and Biographical Introduction, by W. BENHAM.

VIRGIL'S WORKS. Rendered into English Prose. With Introductions, Notes, Analysis, and Index, by J. LONSDALE, M.A., and S. LEE, M.A.

HORACE. Rendered into English Prose. With running Analysis, Introduction, and Notes, by J. LONSDALE, M.A., and S. LEE, M.A.

MILTON'S POETICAL WORKS. Edited, with Introductions, &c., by Professor MASSON.

Published every Thursday, price 6d.; Monthly parts, 2s. and 2s. 6d., Half-Yearly Volumes, 15s.

NATURE:

AN ILLUSTRATED JOURNAL OF SCIENCE.

NATURE expounds in a popular and yet authentic manner, the GRAND RESULTS OF SCIENTIFIC RESEARCH, discussing the most recent scientific discoveries, and pointing out the bearing of Science upon civilisation and progress, and its claims to a more general recognition, as well as to a higher place in the educational system of the country.

It contains original articles on all subjects within the domain of Science; Reviews setting forth the nature and value of recent Scientific Works; Correspondence Columns, forming a medium of Scientific discussion and of intercommunication among the most distinguished men of Science; Serial Columns, giving the gist of the most important papers appearing in Scientific Journals, both Home and Foreign; Transactions of the principal Scientific Societies and Academies of the World, Notes, &c.

In Schools where Science is included in the regular course of studies, this paper will be most acceptable, as it tells what is doing in Science all over the world, is popular without lowering the standard of Science, and by it a vast amount of information is brought within a small compass, and students are directed to the best sources for what they need. The various questions connected with Science teaching in schools are also fully discussed, and the best methods of teaching are indicated.